Philosophy of Science

The book is a translation of the second edition of a much-used and research-based Chinese textbook. As a succinct and issue-based introduction to the Western philosophy of science, the book brings eight focal issues in the field to the fore and augments each topic by incorporating Chinese perspectives.

Followed by an overview of the historical framework and logical underpinnings of philosophy of science, the book thoroughly discusses eight issues in the discipline: (1) the criteria of cognitive meaning, (2) induction and confirmation, (3) scientific explanation, (4) theories of scientific growth, (5) the demarcation between science and pseudoscience, (6) scientific realism and empiricism, (7) the philosophy of scientific experimentation, and (8) science and value. Not confined to Western mainstream discourse in this field, the book also introduces voices of Chinese philosophers and adopts a stance that productively combines logical empiricism and Kuhnianism, both of which tend to be covered in less detail by many English-language textbooks. In the final chapter the author offers a prognosis regarding the future of the discipline based on recent trends.

This book will be of value to students who study philosophy of science and hope to gain a better understanding of science and technology.

Wei Wang is Professor of philosophy of science in the Department of History of Science, Tsinghua University, China.

China Perspectives

The *China Perspectives* series focuses on translating and publishing works by leading Chinese scholars, writing about both global topics and China-related themes. It covers Humanities & Social Sciences, Education, Media and Psychology, as well as many interdisciplinary themes.

This is the first time any of these books have been published in English for international readers. The series aims to put forward a Chinese perspective, give insights into cutting-edge academic thinking in China, and inspire researchers globally.

To submit proposals, please contact the Taylor & Francis Publisher for China Publishing Programme, Lian Sun (Lian.Sun@informa.com).

Titles in philosophy currently include:

The Principles of New Ethics II
Normative Ethics I
Wang Haiming

The Principles of New Ethics III
Normative Ethics II
Wang Haiming

The Principles of New Ethics IV
Virtue Ethics
Wang Haiming

The Metaphysics of Philosophical Daoism
Kai Zheng

Philosophy of Science
An Introduction to the Central Issues
Wei Wang

For more information, please visit
www.routledge.com/China-Perspectives/book-series/CPH

Philosophy of Science

An Introduction to the Central Issues

Wei Wang

First published 2021
by Routledge
2 Park Square, Milton Park, Abingdon, Oxon OX14 4RN

and by Routledge
52 Vanderbilt Avenue, New York, NY 10017

Routledge is an imprint of the Taylor & Francis Group, an informa business

English Version by permission of Tsinghua University Press.

British Library Cataloguing-in-Publication Data
A catalogue record for this book is available from the British Library

Library of Congress Cataloging-in-Publication Data
A catalog record has been requested for this book

ISBN: 978-1-138-84081-2 (hbk)
ISBN: 978-1-315-72823-0 (ebk)

Typeset in Times New Roman
by Newgen Publishing UK

Contents

Figures

Tables

Preface to the English version

Many thanks are due to Routledge and the Taylor & Francis Group for publishing the English version of this book. The Chinese version was first published in 2004, with the second edition following in 2013. The Chinese version, as the first issue-centered textbook on philosophy of science published in Chinese, has been of some influence within philosophical circles in China. But what need is there to publish an English version when there are already so many other textbooks on philosophy of science available in English-speaking countries?

My own work can contribute to the literature in English by providing two kinds of diversity. First, as the name of the series *China Perspectives* suggests, this book attempts to provide some perspectives from the Chinese context by mentioning numerous noteworthy Chinese philosophers of science and by quoting a range of Chinese philosophical literature, both ancient and modern. Second, I myself am a Kuhnian, or at least – and in much the same vein as Professor John Earman dubbing himself "a distant student of Carnap and a close student of Kuhn" – I try to combine logical empiricism and Kuhnianism in a productive way. As I have seen, however, a host of English textbooks have taken a relatively negative attitude towards either Kuhn or logical empiricists, so I hope my own perspective, and especially my treatment of relativism as the biggest issue in the general philosophy of science, can offer a helpful addition to the conversation.

Rebecca Willford at Routledge and Eliza Lai and Qiaozheng Wang at Tsinghua University Press have all provided me with great assistance in making this publication a reality.

The translation of this book was greatly supported by my Ph.D. student Jorge Luis García Rodríguez, who recently received his doctorate in the summer of 2020 and will soon begin teaching at Inner Mongolia University. Ryan Pino, also the editor of my first book in English, *Explanation Laws and Causation*, kindly polished this book as well. He received his M.A. from the School of Philosophy at Fudan University in Shanghai, and he is not only excellent at philosophy, but he is also experienced at editing, so his polishing has even improved the book. He is now a Ph.D. student in Religious Studies at Harvard University, and I sincerely wish him a brilliant academic career in the future. Of course, I take full responsibility for any errors that might remain in the book.

Preface to the Chinese second edition

The first edition of *Philosophy of Science: An Introduction to the Central Issues* was published in 2004. At that time, I had been teaching the undergraduate course "Modern Western Philosophy of Science" for three years. With a little experience, I finally wrote the first edition. After the first edition of the book was completed, I then had the honor of visiting Harvard University in the 2004–2005 academic year, the University of Pittsburgh in the 2005–2006 academic year, and the Institute of History and Philosophy of Science and Technology in France in the fall semester of 2008 to improve my academic competence. 2011 also marked the tenth anniversary of my return from the Chinese University of Hong Kong to Tsinghua University, and I have accumulated new experiences in teaching during that time. Therefore, I am determined to revise the book, expand the content, correct mistakes and oversights, and try to provide a better and updated version for the academic community in China.

There are three main updates to the second edition. First, chapters such as "The nature of laws of nature," "A philosophical analysis of the concept of reduction," and "Philosophy of scientific experimentation" have been added. These topics also summarize some of the topics on which my own research has focused in previous years. Second, some mistakes or oversights in the original edition have been corrected, and some content has been updated. Especially since I received my Ph.D. from the Chinese University of Hong Kong, the text and translation of the original version follow conventions more common in Hong Kong and Taiwan, so the new version uses more mainland Chinese conventions. Third, to stay more in line with the book's focus on the general philosophy of science, one original chapter titled "Philosophy of social science" has been deleted, though I hope to include it in my future monograph on philosophy of specific sciences.

The reprint of this book benefited greatly from my visit to the University of Pittsburgh. Professors John Earman, John Norton, Sandra Mitchell, Merrilee Salmon, Adolf Grünbaum, Anil Gupta, and others from the University of Pittsburgh, as well as Professor Clark Glymour at Carnegie Mellon University, all offered me academic advice and strong support. I thank them very much!

My former M.Phil. student Lei Jiang proofread the book carefully and put forward many very good opinions, especially contributing to the reprint. Teachers desire to see their students succeed in their studies and even surpass themselves, and I am happy to say he is now in the Department of History and Philosophy of Science at the University of Pittsburgh. I also anticipate the second edition to be more accessible for students, as well as more accurate and up to date.

Of course, as they say, "There is no end to learning." Because of the limitations of my own academic ability, there may still be many mistakes in this improved and revised version. I sincerely welcome colleagues and students to criticize and correct my errors.

On Tsinghua's campus, September 2011

Preface to the Chinese first edition

Since 2001, I have been teaching an elective course once a year entitled "Modern Western Philosophy of Science" for undergraduates at Tsinghua University, so this course is now in its third year. Tsinghua students' literacy in science and engineering is already very good, and I wholeheartedly hope to contribute to their development and not to mislead them. Therefore, in supplementing their academic background, I have tried my best to introduce the knowledge of Western philosophy of science to them.

Most of the traditional works on philosophy of science in China focus on philosophers, listing famous scholars and schools throughout history one by one. Wanting to make some innovations and not to talk about philosophy merely as a list of philosophers' viewpoints or anecdotes, I have tried instead to focus on the central issues in philosophy of science in order to introduce these to students. After several attempts over the years, I have accumulated some experience in this regard and achieved good teaching results.

Nevertheless, in the teaching process, I came to feel that there was a lack of issue-centered and comprehensive research-based textbooks on philosophy of science available in China, so with youth and vigor, I bravely set out to prepare a textbook on philosophy of science that is more in line with those available in English-speaking countries. As soon as I started to write, however, I began to understand the adage, "When it comes to applying book learning, one wishes one had read more." My understanding of many issues in philosophy of science was still not deep, and I did not know as much about the literature as I had thought. I could only comfort myself from time to time with knowing this: Writing is a process of self-learning. Finally, I got up the courage to finish the whole book.

Focusing on the issues, this book provides a comprehensive and in-depth study of eight major issues in Western philosophy of science: the criteria of cognitive meaning, induction and confirmation, scientific explanation models and their problems, models of the growth of science, the demarcation between science and pseudoscience, scientific realism, science and value, and philosophy of social science. This book is suitable for graduates and undergraduates in philosophy of science, though college students majoring in science and technology can also increase their understanding of science through this book.

There are undoubtedly many mistakes within. Fortunately, I have always believed in the creed of clarity first, that the writing style should be simple and clear, not perfunctory or abstruse. Therefore, if there are any mistakes in this book, they should be like the saying "The eclipse of the Sun and the Moon can be seen by everyone." I hope that when it comes to later generations, "Everyone changes it." To paraphrase Heidegger, this book is "Wege – nicht Werke." I sincerely hope that the publication of this book can play the role of "throwing away a brick in order to get a gem." Likewise, younger generations of philosophers of science can criticize the mistakes and oversights of this book and make even greater achievements.

In 1993, I entered the Institute of Science and Technology and Society at Tsinghua University and studied for a master's degree with Professor Shi-qi Kou. With the guidance and support of Shi-qi Kou, Xiao-xuan Zeng, Yuan-liang Liu, Hui-hua Yao, and other professors, I received foundational training in philosophy of science and had the opportunity to go to Hong Kong for further study.

From the beginning of 1997 to the beginning of 2001, I then studied for a doctorate in the Department of Philosophy at the Chinese University of Hong Kong and further received systematic and formal training in Western philosophy. Hsiu-hwang Ho, Yuan-kang Shi, Tze-wan Kwan, Tien-ming Lee, Kai-yee Wong, Te Chen, and other teachers in the Department of Philosophy have all cultivated and taken good care of me, which has greatly improved my academic studies. Furthermore, Pow-chung Chow, Yat-tung Chan, Siu-fu Tang, Wai-sang Tang, and other classmates and friends have constantly inspired me in my studies. The flourishing and bustling city of Hong Kong and the beautiful and peaceful campus of CUHK still stir me with nostalgia.

After I returned to Tsinghua University to teach in 2001, Professor Guo-ping Zeng created a good working environment and actively sought out academic resources for me. The publication of this book was also strongly supported by the Institute of Science, Technology and Society at Tsinghua University. Ms. Jing Zhou and other editors at Tsinghua University Press have worked hard to see this book through to publication. Here, I would finally like to express my wholehearted thanks to all those who have cared for me.

The publication of this book is supported by the Beijing Social Sciences Publication Fund for Theoretical Works and the Discipline Development Fund of the School of Humanities and Social Sciences at Tsinghua University.

On Tsinghua's campus, December 2003

1 Introduction

Starting in the twentieth century, philosophy of science has become a rapidly developing branch of philosophy in the West. As a consequence, many leading universities have established specialized departments for philosophy of science (e.g., Department of History and Philosophy of Science, Department of Logic and Philosophy of Science, etc.), international academic networks have formed around related subjects (e.g., the Philosophy of Science Association in America and the European Philosophy of Science Association in Europe), and specialist academic journals have appeared (e.g., *Philosophy of Science*, *British Journal for the Philosophy of Science*, etc.).

Western philosophy of science has also spread extensively in the Chinese academic world, particularly since the 1980s, with China going so far as to establish a specialized committee for philosophy of science called the Chinese Philosophy of Science Association (CPSA), which is subordinated to the Chinese Society for Dialectics of Nature. Consequently, CPSA members have done extensive research in this field and have obtained significant results in the past three or four decades. In light of this growth in the field, both in the West and in China, this book is conceived as a comprehensive introduction to the main issues in Western philosophy of science.

What is philosophy of science?

If we want to know what philosophy of science is, we should start by first asking ourselves the following question: What is philosophy of science *not*? According to E. D. Klemke (1998), philosophy of science, first of all, is not history of science. History of science examines the development of science, as well as great contributions made by scientists. In spite of the close connection between philosophy of science and history of science, however, the former is not concerned with historical research.

Second, philosophy of science is neither cosmology nor natural philosophy, for these two are concerned with issues like the infinite divisibility of matter, the ultimate end of the universe, and so on. Certainly, philosophy of science entails knowledge about the natural world, but the class of issues described above remains more within the scope of scientific research.

Third, philosophy of science is neither sociology of science nor psychology of science, which are concerned with social phenomena related to science and scientists' mental processes, respectively. For instance, the effects of the Strategy of Rejuvenating China through Science and Education or of Albert Einstein's psychological processes during his debate with the Copenhagen School would be issues addressed within these particular disciplines. Although philosophy of science is certainly nurtured and informed by the research conducted in these fields, it does not primarily consist of such kinds of empirical research.

Finally, philosophy of science is not science itself. The aim of research in the natural sciences is the discovery (or invention) of laws of nature by means of mathematical formulation and experimentation. Philosophy of science, however, is not primarily concerned with those previously mentioned fields of research. Instead, it is mainly concerned with issues like the logical form of laws of nature, the necessity of the scientific method, and so on. The object of research of both – that is, philosophy of science and science itself – as well as their mutual relationship, can be represented as follows:

At this point, we might finally ask: What then is philosophy of science? Klemke has given us a helpful preliminary definition: "the aim of philosophy of science is the understanding of the significance, methodology, and logical structure of science by means of logical as well as methodological analysis of its purpose (object), methods, norms, concepts, laws and theory in general" (Klemke 1998, pp. 19–20). According to this definition, we can infer that philosophy of science takes science itself as its research object and an enhanced understanding of science as its aim.

In fact, philosophy of science, like any other practical activity of mankind, such as science or art, is embedded in a continuous process of change. In its earliest period, the main research method was logical analysis; at a later stage of development, however, it also began to incorporate historical methods, as well as sociological research. Other areas of research within the broader discipline of philosophy of science, including scientific experimentation and scientific modeling, have also experienced rapid development and change. In light of this, the author's aim is not to offer a precise definition of philosophy of science; instead, the aim is to convey to the reader a deeper and more comprehensive understanding of its essence through the exposition of its main issues and research methods.

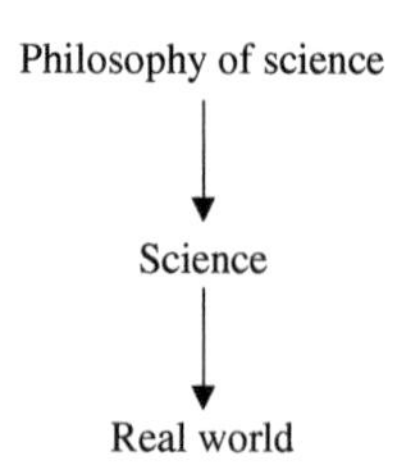

Figure 1.1 Relationship between philosophy of science, science, and the real world

From the personages to the issues

In the early part of the 1930s, the Chinese scholar Tscha Hung (Qian Hong, 1902–1992) began his doctoral work at the University of Vienna and became one of the early members of the Vienna Circle. After his return to China, he took charge of the Institute of Foreign Philosophy at Peking University. Although his research was interrupted during the tumultuous years of the Cultural Revolution, with the eventual introduction of the Reform and Opening-Up Policy, he was finally able to edit the two-volume work *Logical Positivism* (1982) and publish both *The Philosophy of the Vienna Circle* (1989) and *On Logical Positivism* (1999), along with other titles introducing the Vienna Circle and its philosophy to the Chinese people.

Moving into the 1980s, the Chinese philosophical community thus embarked on a more detailed and comprehensive re-elaboration of Western philosophy of science. Even in those early years of the 1980s, many excellent textbooks of philosophy of science appeared in China. A few notable examples include Tian-ji Jiang's *Modern Western Philosophy of Science* (1984), *A Commentary on Modern Western Philosophy of Science* (1987) edited by Wei-guang Shu and Ren-zong Qiu, and *Western Philosophy of Science* (1987) compiled by Ji-song Xia and Fei-feng Shen.

From the 1990s onward, more titles appeared, such as *The Frontiers and Developments of Philosophy of Science* (1991) edited by Shun-ji Huang and Da-chun Liu, *Introduction to the Philosophy of Science* (1996) by Zheng-kun Yin and Ren-zong Qiu, *General Theory of Philosophy of Science* (1998) edited by Da-chun Liu, and Liu's later monograph *Introduction to the Philosophy of Science and Technology* (2000). Also of note around the turn of the millennium were *Towards the 21st Century Philosophy of Science* (2000) edited by Gui-chun Guo and *Course of Philosophy of Science and Technology* (2000) by Wei-tong Sheng.

This body of work contributed greatly to the popularization of philosophy of science and to the promotion of other related fields of investigation. In most of these writings, the exposition revolves around the main personages of each period, arranged and treated chronologically. In other words, they introduce the main philosophers of science and their schools of thought during each of the following historical periods: logical atomism (with representative figures being Bertrand Russell and Ludwig Wittgenstein), logical positivism (Morris Schlick and Rudolf Carnap), logical empiricism (Hans Reichenbach and Carl Hempel), critical rationalism (Karl Popper and Imre Lakatos), historicism (Thomas Kuhn), postmodernism (Paul Feyerabend), and scientific realism (Dudley Shapere and Mario Bunge).

While such books have unquestionable value in conveying to readers the conclusions of these various philosophers and schools of thought, this approach also has its shortcomings. In the first place, it does not take into account those important philosophers of science who made great contributions but whose ideas never crystallized into a particular tradition or school of thought (e.g., Nelson Goodman, Alonzo Church, etc.). Second, this way of writing presents philosophy of science merely as an enumeration of points of view, as if its developments – unlike those in the natural sciences, for instance – have been arbitrary or void of any internal reasoning. Consequently, many people consider philosophical inquiry in this area as hardly more than the nonsensical rumination of scholars. Actually, in China there are many scholars who are overly enthusiastic about establishing their own philosophical schools and becoming famous, and this situation has become a serious obstacle for the development of philosophy of science as an academic discipline. Third, writing that focuses on great philosophers allows readers to keep track of the world's more advanced philosophical trends, but it fails to help philosophical circles within China to innovate and develop their own research. After all, the reason why particular philosophers or schools of thought have triumphed over others is generally because they have succeeded in solving difficult issues or have introduced new and illuminating issues within philosophy of science. Therefore, from the perspective of disciplinary development, the issues of philosophy are prior to the philosophers themselves, not the other way around. In order to deeply understand the profound implications of these issues, particular attention must be drawn to the issues themselves, in addition to the study of those great philosophers.

At the present time, a majority of the textbooks about philosophy of science circulating internationally are centered on the major issues of the field. Accordingly, this book also attempts to describe the main issues of philosophy of science in an issue-centered way. Although China has brought forth few books that focus on the issues, most authors continue to be too driven by the Chinese research program of Marxism, with little concern for international integration. With that said, some monographs have appeared in China focusing on certain issues within philosophy of science, such as Xiao-ping Chen's *Inductive Logic and Inductive Paradox* (1994), Jian Chen's *Demarcation in Science* (1997), and Gui-chun Guo's *Course of Scientific Realism* (2001). However, these books, by limiting their focus to specific issues, offer no comprehensive exposition of the central issues in philosophy of science. On account of this, the current volume is intended to show the range of content dealt with in philosophy of science and the beauty of the analytic method through an issue-centered exposition.

Certainly, this kind of exposition may leave beginners feeling lost in their study of the various issues, causing them to lose sight of the overall historical trajectory of philosophy of science. For this reason, this book also includes a chapter entitled "Historical introduction," which outlines philosophy of science according to its historical development. Utilizing the notion of a "hermeneutical circle," the author hopes this additional chapter on history will help to familiarize readers with the main figures and schools in philosophy of science, so that after having delved into the study of concrete issues, they might acquire a better understanding of the field of philosophy of science as a whole.

The central issues of philosophy of science

What are the central issues of philosophy of science? In the third edition of his book *Introductory Readings in the Philosophy of Science* (1998), which is one of the most widely used textbooks in philosophy of science worldwide, Klemke presents 17 main issues in philosophy of science. These 17 issues and their pertaining questions are as follows:

(1) *Formal science*. In which sense is formal science (mainly mathematics and logic) a science? How can we have knowledge about the truth of logical and mathematical propositions? What is the relation between formal science and experimental sciences such as physics and biology?
(2) *Scientific description*. What is a sufficient scientific description? What is the logical structure of concept formation in scientific description?
(3) *Scientific explanation*. What is scientific explanation? How many models of scientific explanation are there? What is the relation between scientific explanation and science itself?
(4) *Prediction*. Why can science make successful predictions? What is the relation between scientific prediction and explanation?
(5) *Causality and scientific laws*. What is the nature of the relation between causality and scientific laws? Are there non-causal scientific laws?
(6) *Theory, models, and the scientific system*. What is a scientific theory? What is the relation between scientific theory and scientific law? What is a scientific model? What is the role of models in science?
(7) *Determinism*. What are the implications of determinism in science? Is determinism true?
(8) *Philosophical problems of physics*. Has the theory of relativity introduced a subjective component into science? Has quantum mechanics overthrown determinism?
(9) *Philosophical problems of biology and psychology*. Are these two disciplines really different? Can they finally be reduced to physics?
(10) *Social sciences*. Are the social sciences real scientific disciplines? What is the difference between natural and social sciences?
(11) *History*. Is history a science? Does the scope of history include general laws?
(12) *Reductionism and the unity of science*. Is it possible to reduce all sciences to some fundamental science (e.g., physics) so that "the unity of science" can be achieved?
(13) *The extension of science*. Sometimes scientists make metaphysical assertions, like those regarding the eventual heat death of the universe, but can science make these kinds of assertions?
(14) *Science and value*. Is science value neutral? What is the relation between science and value?
(15) *Science and religion*. Can the conclusions of science affect religious beliefs?

(16) *Science and culture*. What is the relation between them?
(17) *The limits of science*. Does scientific knowledge have any limitations? What is the standard of limitation? (Klemke 1998, pp. 22–23)

From this long list, Klemke chooses to discuss six issues in particular: (1) science and pseudoscience, (2) natural science and social science, (3) laws and explanation, (4) theory and observation, (5) confirmation and acceptance, and (6) science and value.

In another popular textbook edited by Martin Curd and J. A. Cover, *Philosophy of Science: The Central Issues* (1998), the editors tackle nine issues in philosophy of science: (1) science and pseudoscience; (2) rationality, objectivity, and value in science; (3) the Duhem–Quine thesis and underdetermination; (4) induction, prediction, and evidence; (5) verification and relevance, using a Bayesian approach; (6) models of scientific explanation; (7) laws of nature; (8) theory reduction; and (9) empiricism and scientific realism.

The issues addressed in the current volume have mainly been selected with reference to these two aforementioned books, *Introductory Readings in the Philosophy of Science* and *Philosophy of Science: The Central Issues*. However, the author has added a new issue in the second edition: the philosophy of scientific experimentation.

That being said, it is necessary to remark that these two referenced books are focused on the issues of general philosophy of science. Most of the 17 issues listed by Klemke are about the general philosophy of science, as the discussion of scientific explanation models is applicable to physics, chemistry, and all the other branches of the natural sciences, for instance. Thus, among these 17 issues, we can see that only issues (8) and (9) – philosophical issues of physics and philosophical issues of biology and psychology – are limited to specific sciences.

In fact, the amount of research conducted in the West on the philosophy of specific sciences is immense. According to the author's estimation, from 1997 to 2000 alone, out of a total of 72 articles published in the *British Journal for the Philosophy of Science*, approximately 27 papers discussed philosophical issues related to specific sciences. In 2000, the American journal *Philosophy of Science* included 36 papers, of which around 16 (almost half) dealt with specific sciences.

At the 1998 biennial meeting of the Philosophy of Science Association (PSA), a total of five main topics were proposed: (1) metaphilosophy and history of philosophy of science; (2) experiment and conceptual transition; (3) philosophy of biology, psychology, and neuroscience; (4) philosophy of physics and chemistry; and (5) philosophy of social sciences. As we can see, among these, there were three main topics – (3), (4), and (5) – related to the philosophy of specific sciences.

During the 2000 PSA biennial meeting, a total of 12 topics were proposed: (1) the metaphysics and methodology of science; (2) the foundations of probability theory; (3) Bayesian methodology; (4) theory dependence and neurobiology of perception; (5) cognition and philosophy of biology; (6) quantum mechanics; (7) statistical mechanics; (8) field theory and relativity; (9) quantum gravity; (10) theory, models, and analogy; (11) science and value; and (12) history and philosophy of science. Among these, there were seven issues related to the philosophy of special sciences, accounting for more than half.

Furthermore, the international ranking of philosophy departments not only includes the general philosophy of science, but also the philosophy of specific sciences, such as the philosophy of physics and the philosophy of biology, as seen in the Philosophical Gourmet Report (c.f. www.philosophicalgourmet.com/breakdown.asp), for instance. Therefore, research in the philosophy of specific sciences has been quite fruitful and holds a very important place in Western philosophy of science. In China, however, the situation is quite different, as most research still emphasizes aspects within the general philosophy of science.

Naturally, doing research on the philosophical issues of specific sciences not only requires knowledge of philosophy, but also demands a deep understanding of the current theories within the particular science under examination. Because of this necessity, the author plans to spend several more years in the preparation of another monograph about the philosophy of specific sciences as a continuation of the current topic. The present book, however, only tackles the central issues wrestled with in the general philosophy of science.

The structure of the book

This book is divided into three parts and has a total of 12 chapters. The first part is introductory, with the first chapter explaining what philosophy of science is, expounding its main issues, and providing a sketch of the overall contents of the book.

The second chapter, "Historical introduction," reviews the main development of philosophy of science from the twentieth century onward. It also introduces the main schools and philosophers of science in chronological order to provide the reader with a historical framework of reference in his/her study of philosophy of science.

The third chapter, "Introduction to logic," briefly explains the basic logical symbols and the corresponding logical rules. It must be noticed that the creation and development of mathematical logic had an immense influence on the rise of Western philosophy of science. Even though historical and sociological methods have been steadily introduced into the study of philosophy of science since the 1960s, the tool of logical analysis remains the most frequently used for research in the field. Hence, it would be extremely difficult, if not impossible, to do any work in philosophy of science without knowledge of logic.

In the second part, the author discusses ten issues in philosophy of science, arranged in the following chapters and sections. In the fourth chapter, "Criteria of cognitive significance," the development of the criteria of cognitive meaning by logical empiricism is discussed. Starting with the principle of testability (including the verification principle, the falsification principle, and the confirmation principle), the principle of translatability (including the requirements of definability and of reducibility) is discussed, and finally there is an explanation of why the strict division between science and metaphysics was abandoned.[1]

In Chapter 5, "Induction and confirmation," the author introduces some inductive methods, expounds Hume's problem of induction, and enumerates several attempts to defend induction. In this chapter we also tackle the problem of confirmation discussed by Carnap, Hempel, Goodman, and others. Beyond this, the author also briefly introduces some recent studies of Bayesianism.

Chapter 6, "Scientific explanation models and their problems," introduces Hempel's covering law models of explanation, such as the Deductive-Nomological (DN) model and the Inductive-Statistical (IS) model. We also discuss the problems of Hempel's explanatory models and the essence of scientific explanation itself.

Two chapters appearing in the Chinese version of this book, "The very nature of laws of nature" (Chapter 7 in the original) and "The philosophical analysis of the concept of reduction" (Chapter 8 in the original), have already been translated into English and can be found in my previous book, also published by Routledge, titled *Explanation, Laws, and Causation* (Wang 2017). Thus, these two chapters are omitted here.

In Chapter 7, "Theories about the growth of scientific knowledge," we examine different theories about the development of science, ranging from the accumulation theory advocated by the logical positivists, Popper's continuous revolution theory, Kuhn's paradigms and incommensurability, and Lakatos' methodology of scientific research programmes, to Feyerabend's anti-credo of "anything goes." At the end of the chapter, the relativistic implications of Kuhn's historicism are briefly discussed.

In Chapter 8, "Demarcation between science and pseudoscience," we discuss the demarcation criteria between science and pseudoscience. These vary from the absolute standard of logicalism to the relative standard of historicism, including postmodernists' dissolution standard and the recent pluralist criterion.

Chapter 9, "Scientific realism," provides a discussion on the problem of whether theoretical terms (e.g., electrons) have real references or whether they are merely instrumental constructions to "save the phenomena" (a notion also explained). The former interpretation has been called scientific realism, while the latter includes constructive empiricism, the Natural Ontological Attitude, and related views.

Chapter 10, "Philosophy of scientific experimentation," both criticizes the theory-oriented view of experimentation in traditional approaches to philosophy of science and expounds the main points of New Experimentalism (e.g., the multiple relations between theory and experiment, that the experiment has its own life, that the experiment creates the phenomena, etc.). This chapter also offers a basic discussion about the sociological analysis of scientific experiments.

In Chapter 11, "Science and values," we analyze the theoretical background of the claim that "science is value free": objectivism. We also introduce some of the progress in related research in philosophy of science, such as the work of Hempel and Kuhn claiming that science is actually value-dependent.

In the third part of this book, Chapter 12, titled "New developments in philosophy of science," some of the recent trends in the "new age" of philosophy of science – such as constructivism, feminism, and postmodernism – are introduced. The author then proposes his own forecast regarding the future philosophy of science.

The significance of philosophy of science

Asking what the significance of X is or whether X has significance at all is not a precise question. Strictly speaking, the correct expression would be: What is the significance of X for someone to do something? For example, for 12th grade students in China, the University Entrance Exam Guide is a significant resource in their preparation for this major examination, but for undergraduates already enrolled at Tsinghua University, it is useless. Beethoven's music may be significant for those in adverse circumstances to keep fighting to achieve spiritual consolation, while it may seem meaningless to moneygrubbers. Thus, when talking about the *significance* of philosophy of science, we should also consider *for whom* it is significant.

Philosophy of science is a theoretical discipline. Therefore, it inherently has theoretical significance. First of all, it is a philosophical reflection about science, so for those dedicated to scientific research, it has guiding significance. Scientists are usually trained extensively in their own disciplines but may spend little time reflecting on the disciplines themselves. For example, scientists may frequently elaborate scientific explanations but know little about the logic of scientific explanation. We use scientific theories and scientific concepts every day, but we often do not understand the nature of scientific theories and concepts. For those especially talented and ingenious scientists, intuitive thinking may be enough to make great scientific achievements, but for the average scientist, some training in philosophy of science may be helpful to engage in science more intelligently and efficiently.

Philosophy of science is also an important component of philosophy. Tian-ji Jiang argues that with the rise of modern science and epistemology as the central issues in philosophical research, philosophy of science, as research and reflection on scientific knowledge, has come to occupy an increasingly important place in philosophy as a whole (Jiang 1984, 1). Thus, it is advisable that all professional philosophers have some knowledge of philosophy of science. Besides that, philosophical research must avoid conceptual vagueness and empty expressions in the same way that philosophy of science emphasizes clarity of expression and logical analysis. Therefore, some training in philosophy of science may be very helpful even for those who devote themselves to other areas of philosophical inquiry.

Finally, a case can be made that philosophy of science even has some significance for our daily lives. The dilemma of modern society is that people have a great deal of pressure in their lives in the pursuit of money and success, yet, at the same time, there are more distractions than ever, making it very easy to lose one's way or develop a kind of tunnel vision. Let us consider an analogy. A psychological experiment has been performed in which a barrier is placed in front of a monkey and a banana is placed on the other side of the barrier. The results have shown that when the banana is placed too far from the barrier, the monkey will circumvent the barrier and easily get the fruit; however, if the banana is placed near the barrier, the monkey will extend its hand through the barrier and try to grab the banana. In the latter case, even if it fails to grab the banana, the monkey will still not know how to circumvent the barrier. If we think about it, a similar sort of scenario often occurs in our own lives and can be described by the phrase "being blinded by lust for money." Hopefully, even seemingly mundane discussions and abstract logical analysis in philosophy of science can serve a more basic human good than simply advancing theory, helping us also to step back for a moment from the noisy world around us and see things a bit more clearly.

Note

1 Rigorously speaking, this issue had been sufficiently discussed by the 1960s and is no longer a central issue in philosophy of science. The reason why the author includes this topic in the first issue is because the change of criteria of cognitive meaning sufficiently demonstrates logical empiricists' technique of logical analysis, and therefore may help the beginner to understand the principal research method in philosophy of science.

2 Historical introduction

From scientific philosophy to philosophy of science

Both of the English terms "scientific philosophy" and "philosophy of science" are translated into the same Chinese term 科学哲学 (*kexue zhexue*). In fact, this reflects the history of the term, as "philosophy of science" evolved from "scientific philosophy." Scientific philosophy originated in the 1920s with the logical positivism represented by the Vienna Circle.[1] In his book *The Rise of Scientific Philosophy*, Reichenbach argues that the speculative philosophies of the nineteenth century, particularly Hegel's philosophy, are unscientific. In contrast, Reichenbach advocates a shift to using scientific methodologies in philosophical research, thus calling this proposed approach "scientific philosophy."

Logical positivists regard speculative philosophy as mere "metaphysics," whereas their new form of philosophy is considered the genuine philosophy. Here, a historical explanation of these terms can be helpful to show the difference. The word "philosophy" is derived from the Greek words "love" (*philo*) and "wisdom" (*sophia*), with the resulting meaning of "love for wisdom," which is always praiseworthy. However, the more specific word "metaphysics" is derived from the words "after" (*meta*) and "physics" (*phusika*). This is partly because, in ancient Greece, Aristotle wrote a special treatise on metaphysics, which followed after his *Physics*. In addition, the latter treatise focuses on more fundamental issues than the one on physics, thus the treatise also came to be known as "after physics" or *Metaphysics*. As for the Chinese word for metaphysics, 形而上学 (*xing er shang xue*), it can be traced back to the ancient phrase "Metaphysics is called Way (*Dao*); while physics is called Implementation" (*The Book of Changes*, "Hsi Tzu").

With these distinctions in mind, logical positivists have historically opposed speculative, or metaphysical, philosophy and instead advocated scientific – or, one might say, empirical – philosophy. Since the origins of this school in the twentieth century, however, they have become more concerned with the philosophical issues of science itself, not simply with doing philosophy scientifically. Hence, the term "scientific philosophy" finally evolved into the current term "philosophy of science."

From a historical perspective, the development of "philosophy of science," roughly speaking, includes the stages of logical positivism, critical rationalism, historicism, postmodernism, and the current "new wave." These stages, along with some of the key figures associated with each, can be listed as follows:

Logicalism	Logical atomism: Bertrand Russell, Ludwig Wittgenstein, etc. Logical positivism: Moritz Schlick, Rudolf Carnap, etc. Logical empiricism: Hans Reichenbach, Carl Hempel, etc. Critical rationalism: Karl Popper, Imre Lakatos
Historicism (in the broad sense)	Historicism (in the narrow sense): Thomas Kuhn, etc. Postmodernism: Paul Feyerabend, etc.
New wave	Larry Laudan, Dudley Shapere, etc.

The term "logicalism," though the author's invention, refers broadly to the four schools within the first row: logical atomism, logical positivism, logical empiricism, and critical rationalism. All of these are attempts to study the structure of science by means of logical analysis; this is why the author places all of these within the broader category of "logicalism."[2] Because logicalists wish to offer a general account of the logical structure of science, but deny its historical context, their approach, while rendering many outstanding achievements, still contains numerous difficulties.

The schools in the second row are conceptually divergent in many respects, but they all stem from Kuhn's historicism. Even more basically, they all emphasize the role of historical factors in scientific research. For these reasons, the author classifies them under the category of "historicism," understood in a broad sense. Historicists respect the real history of science and thus adopt some of the traditional methods of history and sociology. Eventually, this approach of historicism came to replace logicalism as the mainstream within philosophy of science; however, because historicists generally sought to eliminate the absolute standard employed by logicalists, the problematic result for scientific knowledge was a certain level of relativism and irrationalism.

Because of the problems resulting from both of these aforementioned approaches, the new wave of philosophy of science, listed in the above chart's third row, aims to reconcile logicalism and historicism by seeking a middle way for philosophy of science. This more balanced attitude is the predominating tendency in Western philosophy of science at the current time. Worth noting is that, in China, Larry Laudan and Dudley Shapere are widely regarded as the key representatives of this new wave.

Above is a very rough sketch of the broad historical developments within philosophy of science. In what follows, the book will briefly introduce each of these schools in a bit more detail.

Logical atomism

As mentioned earlier, modern philosophy of science originated with the Vienna Circle, the dominant approach of which is also called "logical positivism," though many members of the Vienna Circle themselves preferred the term "logical empiricism." As the term suggests, logical positivism is composed of two primary components: "logic" and "positivism." Thus, in order to fully understand the meaning of logical positivism, we must introduce the rise of modern logic and the precursors of logical positivism, namely logical atomism and positivism.

The founder of modern logic was the German philosopher and logician Gottlob Frege (1848–1925). Born in Wismar, Germany on November 8, 1848, Frege entered the University of Jena in 1869, but transferred to the University of Göttingen in 1871, receiving his Ph.D. in mathematics in 1873. In 1874, he became a lecturer at the University of Jena and was promoted to Extraordinarius Professor in 1879 and Honorary Professor in 1896, finally retiring in 1917 and passing away in 1925. His most representative writings include *Begriffsschrift* (*Concept Notation*, 1879), *Die Grundlagen der Arithmetik* (*The Foundations of Arithmetic*, 1884), and the two-volume work *Grundgesetze der Arithmetik* (*The Basics Laws of Arithmetic*, 1893/1903) (Zalta, 2019).

Frege is seen as the father of modern logic because he created the basic framework of symbolic logic. In setting out to establish the foundation of mathematics, he realized that ordinary language is neither sufficiently strict nor precise, so he wished to construct a "formal language of pure thought." Accordingly, he invented the system of first-order logic. In his book *Begriffsschrift*, Frege proposes the expression of propositional logic in the form of an axiomatic system and also invents symbolic language, introducing two quantified variables, *all* and *exist*, and thus setting the foundation of predicate logic. In his *Grundlagen*, Frege first proposes that all mathematical knowledge could be ultimately reduced to logic, or the so-called thesis of "logicism" in the philosophy of mathematics. Beyond these groundbreaking proposals, he also made great contributions to analytic philosophy.

Due to the depth of his thought and the cryptic style of his writings, however, Frege remained relatively unknown during his lifetime. Despite this, his works later garnered much attention and deeply influenced a number of great philosophers, such as Bertrand Russell and Ludwig Wittgenstein. Thus, he eventually came to be widely recognized as "the father of modern logic."

Moving on from Frege, the main figures associated with logical atomism are Russell and Wittgenstein. Bertrand Russell (1872–1970) was born into a noble family, his grandfather having served as the Prime Minister of Britain. In 1890, he entered Trinity College, Cambridge, where he attended seminars on mathematics conducted by Alfred North Whitehead (1861–1947), who had a profound influence on his thinking. From 1910 to 1913, he wrote *Principia Mathematica* in collaboration with Whitehead, and this important work became a milestone in the field of modern mathematical logic. Among Russell's most influential works are also *The Problems of Philosophy* (1912) and *The Analysis of Matter* (1927).

In his day, Russell was not only a great logician and philosopher, but he was also an outstanding writer and highly engaged social activist, even being sent to prison for half a year in 1918 due to his outspoken opposition to the First World War. During the Second World War, he also became a representative figure of the nuclear disarmament movement. As for his popular writings, Russell was prodigious, with non-technical works including *History of Western Philosophy*, *Wisdom of the West*, *Why I Am Not a Christian*, *Marriage and Morals*, and *Power*, among others. Because of his popularly and profundity, it is not surprising that, in 1959, he received the Nobel Prize for Literature in recognition of the beauty of his English writing. As for his relationship with China, he visited the country from October 1920 to July 1921, taught at Peking University, and published *The Problem of China* in 1922. However, his philosophy met with little positive response in China, as many Chinese intellectuals were not satisfied with Russell's proposed solutions to the problems of China. Thus, his visit ended in disappointment.

Professionally, Russell made many contributions to philosophy on a variety of topics. In his paper "On Denoting" (1905), he proposed a "theory of descriptions," which became a paradigm for analyzing ordinary language by means of logical methods. His proposal on this score was based on the fact that ordinary language is often vague in meaning and unclear in structure. For example, he pointed out that the assertion "The present king of France is bald" seems to be false because France is now a republic with no king of which to speak. However, this sentence's negation, "The present king of France is not bald," also seems to be false for the very same reason. As a result, neither the first sentence nor its negation can be true. What, then, is at the root of this problem?

In order to resolve the issue, Russell reformulated the first sentence in a more precise form: "There is one and only one person who is the present king of France, and that person is bald." The logical form of the statement is as follows (let K be "being the present King of France" and B be "being bald"):

$$\exists x(Kx \land \forall y(Ky \rightarrow (x=y)) \land Bx)$$

Thus, through logical reformulation we can clearly see that the problem arises from the fact that the clause "There is one and only one person who is the present king of France" is a false statement, regardless of whether or not that person is bald. Therefore, the negation of this proposition is not simply "The present king of France is not bald," but is rather a disjunctive statement: "Either there is no king of France, or there is more than one king in France, or the king of France is not bald."

Russell also raised the famous "Russell's paradox," which, in turn, caused a third crisis in the history of mathematics. The Russell's paradox can be formulated as the following question: "Is the set of all sets that are not members of themselves a member of itself?" As is commonly known in mathematics and logic, there are some sets that can be members of themselves, such as the set of all "nouns," including the words "students," "Tsinghua University," "nouns," and so on. Other sets, however, are not members of themselves. For example, the set of all "verbs" includes the words "walk," "run," "jump," etc., while the word "verb" itself is a "noun" and is therefore not a member of itself. The set of all sets that are not members of themselves can be represented in the form $A=\{X| X \notin X\}$. Thus, there are two possibilities concerning A. If A is a member of itself, then by definition it must not be a member of itself. Similarly, if A is not a member of itself, then by definition it must be a member of itself. This is evidently a paradox. The Russell's paradox thus prompted a great deal of research on the foundation and accuracy of both logic and mathematics.

Important for us, in the philosophy of mathematics, Russell supported Frege's view of logicism, according to which all mathematical truths are essentially logical truths and all mathematical knowledge can be reduced to logic. In a lecture delivered in 1918, Russell first proposed his conception of "logical atomism." In this philosophical conception, he postulated that the world is constituted of logical atoms. So, for example, a full description of Aristotle can be gained from the combination of indivisible atomic facts, such as "Aristotle was a student of Plato," "Aristotle died in 322 BC," and so on. Ideas such as this that formed the foundation of Russell's logical atomism were then further developed by his student Ludwig Wittgenstein.

Ludwig Wittgenstein (1889–1951) was born in Vienna into a wealthy Jewish family, his father being one of Austria's most powerful steel magnates. Three of his four brothers committed suicide, and Wittgenstein himself was known to have some tendencies toward melancholy. At the beginning of his career, Wittgenstein studied aeronautical engineering, but he later became strongly interested in logic and philosophy. As a result, in 1912, he moved to Trinity College, Cambridge to study philosophy under Russell's supervision.

As mentioned, Wittgenstein had a very unique personality. In 1913, he inherited a huge fortune from his parents, but he soon gave it all away while continuing to live in relative poverty himself. During the First World War, he voluntarily joined the Austro-Hungarian army and served as a machine-gunner. Because of his courageous service, he was commissioned as an officer in 1915. During the war, however, he continued to think deeply about philosophical problems and took copious notes even while he was in the trenches. In 1918, he was interred in a prisoner of war camp for several months, yet he still managed to finish the manuscript of *Tractatus Logico-Philosophicus* during this harrowing time. After his release in 1919, *Tractatus* was finally published in 1921, largely thanks to Russell's assistance.

Tractatus Logico-Philosophicus clearly outlines Wittgenstein's early philosophical position. In this book, he proposes the method of the truth-value table for propositional logic. He also clearly states, for the first time, that analytic propositions, which are necessarily true, are in fact tautologies that offer no description of the world whatsoever. In contrast, synthetic propositions, which describe the world, have empirical significance.

The book also takes the stance of logical atomism. In it, Wittgenstein argues that language consists of propositions, while propositions are truth functions of elementary propositions, and finally elementary propositions are composed of names. In parallel with this, the world is the sum of facts, and facts are composed of states of affairs or atomic facts, which, in turn, are combinations of objects. Hence, elementary propositions and states of affairs have a common logical structure. The following visual represents this relationship between an elementary proposition and a corresponding state of affairs.

To explain, let us say it is the case that today is sunny and all the students in a class are attentively listening to the lesson. This circumstance can be divided into two facts, namely "Today is sunny" and "All the students are attentively listening to the lesson." In turn, the fact "All the students are attentively listening to the lesson" is a combination of states of affairs in the form "X_1 is attentively listening to the lesson," "X_2 is attentively listening to the lesson," etc. These states of affairs cannot be divided into other further facts and are

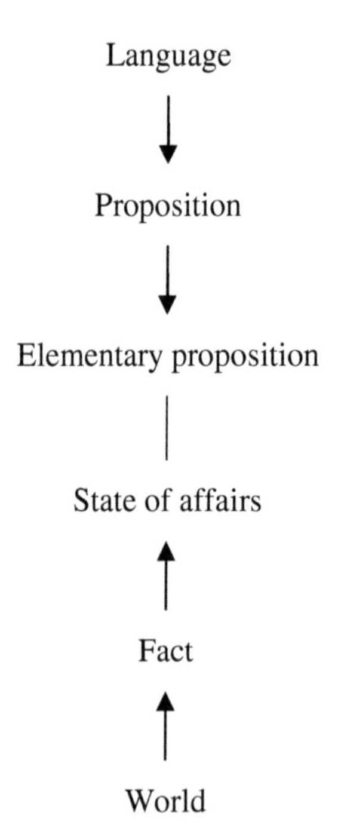

Figure 2.1 Wittgenstein's picture theory in the *Tractatus*

therefore atomic. They are composed of two objects, namely individual persons (such as X_1, X_2, X_3) and "is attentively listening to the lesson."

Because the world and human languages share the same logical structure, Wittgenstein argues, there is a picture correspondence between states of affairs and elementary propositions. Therefore, a state of affairs can be expressed by the elementary propositions "X_1 is attentively listening to the lesson," "X_2 is attentively listening to the lesson," and so on. Accordingly, the proposition "All the students are attentively listening to the lesson" is the conjunction of the elementary propositions above. The two propositions, "Today is sunny" and "All the students are attentively listening to the lesson," combined with the totality of other propositions, constitute human language.

As this demonstrates, Wittgenstein thought that human languages and the world share the same logical structure, so the task of philosophy is to analyze human language by means of logic. This early philosophy of Wittgenstein deeply influenced the logical positivist conceptions of the Vienna Circle.

Interestingly, Wittgenstein believed that his *Tractatus* had solved all the problems of philosophy once and for all, and he consequently turned to other jobs in the years between 1920 and 1926, working as a middle school teacher, then as a gardener, and even designing his sister's house. In 1929, however, he decided to return to Cambridge, where he obtained his Ph.D. with the *Tractatus Logico-Philosophicus* as his dissertation. From there, he resumed his philosophical research.

Wittgenstein gradually gave up his early philosophical position, mainly because of criticisms raised by his friends F. P. Ramsey and Pino Sraffa. It is said that once, while on a train, Wittgenstein insisted that a proposition and what it describes share the same logical form. In response, Sraffa made a Neapolitan gesture of brushing his chin with his fingertips and then asked: "What is the logical form of the action?" Wittgenstein realized the shortcomings of his early position and then turned to criticizing his own work in his later life (Jiang 1998, p. 113).

Wittgenstein passed away in 1951; his last known words were: "Tell them I've had a wonderful life." In 1953, his students compiled his lecture notes and published them under the title *Philosophical Investigations*, containing several important philosophical concepts like "language games" and "family resemblances," while also raising severe criticisms of his earlier philosophy. The book greatly influenced the school of ordinary language at Oxford University, and notable people from this school include John Austin and Peter Strawson. Beyond this, it also had – and, indeed, continues to have – a tremendous influence on many fields, such as philosophy of language, philosophy of mind, and philosophy of mathematics, among others.

In sum, logical atomists did a great deal of fundamental work in mathematical logic; moreover, one of their major and most influential contentions was that the main task of philosophy is to analyze ordinary language by logic. These two achievements did much to lay the foundation for the development of logical positivism.

Logical positivism

Another source of logical positivism other than logical atomism is positivism. Positivism originated with the French philosopher and sociologist Auguste Comte (1798–1857), who proposed the law of three successive stages in the history of ideas: theological, metaphysical, and positive (or scientific). In the theological stage, Comte claimed, people believe in God or supernatural agents in general. In the metaphysical stage, people replace supernatural agents with abstract entities, but they still believe in unverifiable concepts like the unobservable "force" in Newtonian physics. Only after coming to the positive or scientific stage, however, can people reach the highest degree of understanding – in other words, believing only verifiable and measurable phenomena (Ruse 1995, p. 145). Comte suggested that people should only believe in positive knowledge, a suggestion for which he is widely regarded as the founder of positivism.

If we consider Comte to mark the first generation of positivism, then the Viennese physicist Ernst Mach (1838–1916) represents the second. Mach made a host of important contributions to different aspects of physics, such as optics, acoustics, and mechanics. In fact, the eponymously coined term "Mach number," which indicates the ratio of flow velocity past a boundary to the local speed of sound, is one of his great academic achievements in fluid dynamics. In addition to his academic contributions to these fields, however, he was also a great historian and philosopher of science. In his book *The Science of Mechanics*, Mach raises severe criticisms of the Newtonian conception of space and time. For instance, Mach insists that concepts like space and time in physics should be reduced to the positive analysis of sense data. Even Albert Einstein later acknowledged Mach as a great inspiration for his advancement of the theory of relativity.

Building on the positivism of these important figures, Comte and Mach, logical positivism then found its origins in the Circle of Vienna, whose founder and leading figure was Moritz Schlick (1882–1936). Interestingly enough, in 1895, the University of Vienna established a Chair of Inductive Sciences especially for Ernst Mach. Following Mach, the physicist L. E. Boltzmann (1844–1906) held the position from 1902 to 1906. Later, in the year 1922, Schlick took over that very same position. This fact itself suggests the close institutional connection between Mach and what would become the Vienna Circle.

Following Schlick's appointment at the University of Vienna, Hans Hahn, K. Reidermeister, Philipp Frank, and Otto Neurath formed a discussion group called the Schlick Zirkel, which, as the name suggests, took Schlick as the center. Other notable participants included Rudolf Carnap, Karl Menger, Richard von Mises, Friedrich Waismann, Viktor Kraft, Herbert Feigl, and Kurt Gödel. Near the end of the 1920s, Hahn, Neurath, and Carnap established the Vienna Circle on the basis of this Schlick Zirkel. Then, in 1929, Carnap, Neurath, and Hahn, with the assistance of Waismann and Feigl, published a manifesto entitled "The Scientific Conception of the World: The

Vienna Circle," and it was after this publication that the Vienna Circle became known worldwide. Relevant to our current discussion, the term "logical positivism" was first introduced by A. Blumberg and Feigl in 1931 in reference to the philosophical position of the Vienna Circle.[3]

The leading figure of the Vienna Circle, M. Schlick, was born into an aristocratic family on March 4, 1882 in Berlin, Germany. He was originally a physicist and studied under the supervision of Max Planck at the University of Berlin, where he received his Ph.D. in physics in 1904 with a dissertation entitled "On the Reflection of Light in a Non-Homogeneous Medium" ("Über die Reflexion des Lichts in einer inhomogenen Schicht"). In 1917, he wrote the monograph *Space and Time in Modern Physics*, which was the first philosophical interpretation of Einstein's theory of relativity. Schlick also maintained personal correspondence with prominent European physicists, such as Planck and Einstein, as well as the mathematician David Hilbert. The prominent Chinese scholar Qian Hong, who became the head of the Institute of Foreign Philosophy at Peking University, also studied with Schlick from 1928 to 1936 and became a full member of the Vienna Circle. Sadly, on June 22, 1936, Schlick was shot to death by a mentally disordered student at the University of Vienna.

Although Schlick was much like the spiritual leader of the Vienna Circle, Rudolf Carnap also contributed a great deal of technical and formalization work to the new school of logical positivism. Carnap, who was born in 1891 in Ronsdorf, Germany studied mathematics and physics at the University of Jena from 1910 to 1914, but he also attended a number of Frege's courses in mathematical logic. In 1926, he accepted Schlick's invitation and moved to the University of Vienna to be an instructor, and in 1930, he and Reichenbach together started the journal *Erkenntnis*, which became a primary outlet for the Vienna Circle and for analytic philosophy more broadly.

In December of 1935, Carnap emigrated to the United States and taught at the University of Chicago until 1952. In 1954, because of Reichenbach's death, he moved to the University of California at Los Angeles to take his position, remaining at UCLA until his retirement in 1961. His books include notable titles, such as *The Logical Structure of the World* (1928), *The Logical Syntax of Language* (1934), *Introduction to Semantics* (1942), *Formalization of Logic* (1943), *Meaning and Necessity* (1947), *Logical Foundations of Probability* (1950), and *Philosophical Foundations of Physics: An Introduction to the Philosophy of Science* (1966).

With the death of Hans Hahn in 1934, Carnap's exodus to the University of Chicago in 1936, and the death of Schlick in the same year, the Vienna Circle essentially disintegrated. Even worse, in 1938, with the occupation of Vienna by Nazi Germany, the doctrines of the Vienna Circle were declared "reactionary philosophy" and, consequently, were publicly prohibited. Nevertheless, because many of the members of the Vienna Circle emigrated to the United States, the Vienna Circle came to have profound influence in the English-speaking world.

The central doctrines of logical positivism are empiricism and anti-metaphysics, and the members of the Vienna Circle held that the foundation of scientific knowledge should be empirically verifiable sensory experience. They also rejected metaphysics and advocated scientific philosophy, arguing that philosophy should study the meaning of language by means of logical analysis. Accordingly, the task of philosophy according to this school of thought should not be the formulation of propositions or the construction of propositional systems – i.e., theoretical knowledge – for that is the task of science. More precisely, the job of philosophy should be to analyze and clarify the meaning of scientific concepts, hypotheses, and propositions, therefore eliminating the confusion of metaphysics through logical analysis.

What is a meaningful statement? To answer this basic question, logical positivists proposed the principle of verification: The meaning of a statement is the way of its verification. So, for example, the statement "It rained in Beijing on February 14, 2003" is either true or false, and if we check the weather on that particular day in Beijing, the statement can be verified or falsified. In contrast, the statements of metaphysics, such as Hegel's "Reason is Substance, as well as Infinite Power," go beyond the possibility of verification; therefore, they are meaningless pseudo-propositions. Hence, from the perspective of logical positivism, metaphysics has no cognitive significance whatsoever, instead being like the expressions of art or poetry.

Many members of the Vienna Circle, especially Neurath and Carnap, also advocated for the "unity of science," or physicalism, believing that all empirical sciences could be unified. Because the statements of all the empirical sciences can be reduced to physical language, they maintained, it is probable that all empirical sciences can also be reduced to physics, thereby achieving the unification of science. This "unity of science" movement has had tremendous repercussions in Western academic circles, especially impacting the fields of biology and psychology.

With that said, logical positivism was met with a host of intractable problems and challengers. For example, its criteria of cognitive significance have a number of logical difficulties (see Chapter 4 for more details). Furthermore, logical positivism's sharp distinction between theory and observation was severely criticized by Karl Popper, while its analytic–synthetic dichotomy was challenged by W. V. Quine. As a result, by the 1960s, which witnessed the rise of historicism, the heyday of logical positivism had largely run its course.

Logical empiricism

There is no major difference between logical positivism and logical empiricism. The former emphasizes the positive tradition of Comte and Mach, while the latter emphasizes the empiricist tradition of John Locke and David Hume in the United Kingdom. For the purposes of this book, the author mainly identifies "logical positivism" with the Vienna Circle, whereas "logical empiricism" is used to refer to the Berlin Circle, or Berlin Group, and other analytic philosophers of science in the United States and Britain.

The leading figure of the Berlin Circle was Hans Reichenbach (1891–1953). Born to a Jewish family in Hamburg, Germany on September 26, 1891, Reichenbach studied mathematics, physics, and philosophy at Munich, Berlin, and Göttingen, and finally received his Ph.D. from the University of Erlangen in 1915. Among his professors were the philosopher Ernst Cassirer, the mathematician David Hilbert, and the physicist Max Born. From 1917 to 1920, he even worked with Einstein on the theory of relativity, and he later wrote three books on the theory of relativity: *Axiomatization of the Theory of Relativity* (1924, English translation 1969), *From Copernicus to Einstein* (1927, English translation 1942), and *The Philosophy of Space and Time* (1928, English translation 1957). In 1926, he became a professor of philosophy of physics at the University of Berlin, and in 1928, he created the Society for Empirical Philosophy, also known as the Berlin Circle, which included other significant scholars, such as David Hilbert and C. G. Hempel.

Because Adolf Hitler took power in Germany, Reichenbach had to emigrate to Turkey in 1933, where he headed the Department of Philosophy at the University of Istanbul and published his *Theory of Probability* (1935, English translation 1949). In 1938, he moved to the United States and taught at the University of California in Los Angeles. His writings from this later period included *Experience and Prediction* (1938), *Philosophic Foundations of Quantum Mechanics* (1944), *Elements of Symbolic Logic* (1947), *The Rise of Scientific Philosophy* (1951), *Nomological Statements and Admissible Operations* (1954), and *The Direction of Time* (1956).

In his writings, Reichenbach criticized the logical positivist thesis for "verificationism," preferring instead to call himself a "logical empiricist." He offered the relative frequency interpretation of probability theory and provided some justification for the method of induction, and he also offered many philosophical interpretations for modern physics, his position being close to conventionalism.

Another significant figure in the school of logical empiricism, Carl Gustav Hempel was born on January 8, 1905 in Oranienberg, Germany. He studied mathematics with David Hilbert and Edmund Landau and symbolic logic with H. Behmann, both at the University of Göttingen and at the University of Heidelberg. After moving to the University of Berlin in 1925, he met Reichenbach and studied physics with Max Planck and logic with John von Neumann. In 1929, he met Carnap at a congress on scientific philosophy and decided to enter the University of Vienna to study with Schlick and Carnap. In 1934, he received his doctoral degree from the University of Berlin, but due to the rampancy of Nazism, he moved that same year to Brussels, where the philosopher Paul Oppenheim was working. After visiting the University of Chicago in 1937 at Carnap's invitation, Hempel emigrated to the United States in 1939. He taught at Yale University from 1948 to 1955, during which time he published *Fundamentals of Concept Formation in Empirical Science*

(1952), and in 1955 he moved to Princeton University, the school at which he would spend most of his career and publish some of his most relevant works, including *Aspects of Scientific Explanation* (1965) and *Philosophy of Natural Science* (1966). From the years 1976 to 1985, he taught at the University of Pittsburgh. Following retirement, Hempel died in Princeton, New Jersey on November 9, 1997.

Known for having an amiable personality, Hempel was a notoriously open-minded scholar among the logical empiricists. In fact, his paper on the empiricist criteria of cognitive significance announced the end of logical empiricism, and his account of scientific explanation still remains one of the central issues in philosophy of science. In light of this, Chinese scholar Wei-guang Shu highly praised him for having "the merits of both logical empiricism and the school of socio-historicism all combined" (Shu and Qiu 1987, p. 86). Indeed, reflecting Hempel's great contributions to philosophy of science in general, in 2012 the PSA named the award for lifetime scholarly achievement in the philosophy of science the Hempel Award.

Moving away from figures hailing from the European mainland, the main representative of logical empiricism in the United Kingdom would undoubtedly be A. J. Ayer (1910–1989). Ayer was educated at Christ College, Oxford. After graduation, and following the recommendation of philosopher Gilbert Ryle, he went to the University of Vienna to study under Schlick and also attended the meetings of the Vienna Circle. After coming back to the UK, he mainly taught at University College London and Oxford University until his retirement in 1978.

Ayer's most important book, *Language, Truth, and Logic* (1936), expounds comprehensively and in great detail the profound theories of logical empiricism, yet manages to do so in simple language. Because of this, the book has had a tremendous impact on the promotion of logical empiricism in the UK. In addition, Ayer's other writings include *The Foundations of Empirical Knowledge* (1940), *Philosophical Essays* (1954), *The Problem of Knowledge* (1956), *The Concept of Person and Other Essays* (1963), *The Origins of Pragmatism* (1968), *Metaphysics and Common Sense* (1969), *The Central Questions of Philosophy* (1973), and *Philosophy in the Twentieth Century* (1982).

Shifting now across the Atlantic, Willard Van Orman Quine (1908–2000) had a decisive influence on the development of logical empiricism in the United States. Born on June 25, 1908 in Akron, Ohio, Quine attended Harvard University and studied logic under the supervision of A. N. Whitehead. After receiving his Ph.D. in 1932, he taught at Harvard University for most of his career until his retirement in 1979. His long résumé of writings includes the following: *A System of Logistic* (1934), *Mathematical Logic* (1940), *Elementary Logic* (1941), *Methods of Logic* (1950), *From a Logical Point of View* (1953), *Word and Object* (1960), *Set Theory and its Logic* (1963), *The Ways of Paradox and Other Essays* (1966), *Selected Logic Papers* (1966), *Ontological Relativity and Other Essays* (1969), *The Web of Belief* (1970), *Philosophy of Logic*

(1970), *The Roots of Reference* (1974), *Theories and Things* (1981), *The Time of My Life: An Autobiography* (1985), *Pursuit of Truth* (1990), *From Stimulus to Science* (1995), and more. Quine passed away on December 25, 2000.

During the academic year of 1932–1933, Quine went to Europe on a fellowship, and during this time abroad he attended meetings of the Vienna Circle and met Schlick, Neurath, and others. Through this experience, he came to deeply understand logical positivism, and Carnap's philosophy especially exercised an influence on him. However, after the 1940s, Quine began to criticize logical positivism. For example, in his most influential paper, "Two Dogmas of Empiricism," he cast doubt on the two doctrines of logical positivism: the analytic–synthetic dichotomy and the claim that every meaningful statement can be reduced to observational language.

Quine held that analytic propositions of non-logical truths with the form "All As are Bs" – such as "All bachelors are unmarried" – entail the problem of the definition of "synonymy." However, if the very notion of "synonymy" is further analyzed, we find no universal and non-circular definition. Therefore, the dichotomy of "analytic–synthetic" should be abandoned. In addition, Quine suggested holism and opposite reductionism. Following Pierre Duhem, Quine thought that what should be tested and revised by experience should not merely be singular sentences, but the whole knowledge system. As an analogy, he regarded the system of human beliefs as a force field. In this picture, the boundary is experience, whereas the regions near the boundary are empirical generalizations; moving from there, we have the general laws of nature, and at the center are logical truth and ontological commitments. When there is any conflict between our knowledge and experience, we first try to adjust the empirical generalizations, then the universal laws, and finally we might modify the logical knowledge and ontological commitments, if necessary. Thus, Quine concluded, logic and mathematics are fallible, as are other natural sciences. Challenges such as this from Quine were among the main causes of logical positivism's decline during the 1960s.

Quine's own philosophical transformation went from logical empiricism to holism, and then to pragmatism. For that reason, his philosophical position has come to be regarded as new pragmatism or logical pragmatism. Still, others maintain that Quine in fact proposed a transition from a narrow and strict version of logical positivism to a more modest and liberal version of empiricism, or a conventionalistic and pragmatistic empiricism. Thus, he is still regarded as a logical empiricist in many Chinese textbooks on philosophy of science.

Critical rationalism

Moving on from logical empiricism to critical rationalism, the main representative of critical rationalism is undoubtedly Karl Popper (1902–1994). Popper was born in Vienna to Jewish parents, and while he was influenced by his father, a lawyer with keen interests in literature and philosophy who actively engaged in social issues, Popper's mother also inculcated in him a passion for music – so much so that he once contemplated becoming a professional musician and even chose the history of music as the secondary subject on his Ph.D. examination. He began his higher education at the University of Vienna in 1918 and received a Ph.D. in philosophy in 1928. Worth noting is that, although Popper had friendly relationships with many members of the Vienna Circle, he did not share the same philosophical position. In his early years, Popper was very sympathetic to Marxism and even joined the Association of Socialist School Students. Also during his time in Vienna, he encountered Freud's theory of psychoanalysis and served in Alfred Adler's mental health clinic. Popper himself declared that 1919 was a crucial year for him, for in that year astronomers confirmed Einstein's prediction that light coming from stars was bent as it passed the Sun. Einstein's attitude of accepting the possibility of falsification with pleasure also inspired Popper's central thought that the proper scientific attitude should be a critical one.

Due to the rise of Nazism, Popper emigrated to New Zealand in 1937 and taught at the University of Canterbury. In 1946, he moved to the UK and founded the Department of Philosophy, Logic and Scientific Method at the London School of Economics (LSE). Becoming a professor of logic and scientific method in 1949, he remained at LSE until his retirement in 1969. Among his most prominent students was George Soros, the billionaire investor and founder of the Quantum Fund.

Popper's research interests were mainly in philosophy of science and political philosophy, and his representative books include *The Logic of Scientific Discovery* (1934, English translation 1959), *The Open Society and Its Enemies* (1945), *The Poverty of Historicism* (1957), *Conjectures and Refutations* (1963), *Objective Knowledge* (1972), *Unended Quest* (1976), and *The Self and Its Brain* (1977). He died on September 17, 1994 at the age of 92.

Popper objected to the logical positivists' sharp distinction between theory and observation. Instead, he believed that observation is always theory-laden because we do not know how and what to observe without some kind of theoretical orientation. Besides, observational descriptions are often expressed in theoretical terms. For example, the statement "This is a glass of water" seems to be a directly observable fact, but to verify that the chemical formula of the glass of water is H_2O, we may resort to an electrolysis experiment. The result of the electrolysis test makes use of theoretical terms like "oxygen" and "hydrogen," meaning "This is a glass of water" is ultimately a theory-laden statement.

Following this line of thinking, Popper objected to the method of induction, for Hume's problem of induction shows that there is no logical path from limited observations to universal scientific theories (for details, see Chapter 5). In addition, since observations are theory-laden, it would beg the question to induce theories from observations.

Consequently, Popper opposed empiricism and preferred rationalism, and based on the rationalist deductive method, he proposed that the scientific method should be one of hypothesis falsification, with the logical form as follows (T designates theory, O observation):

$$T \rightarrow O, \quad \neg O$$

$$\therefore \neg T$$

In fact, such a method of hypothesis falsification is the *modus tollens* inference in deductive logic, so it is a method of deduction rather than induction. In this way, Popper tried to use falsificationism to eliminate the method of induction in science.

As Popper saw it, there is neither verification nor confirmation in science, but only falsification. After all, scientists boldly propose theories and then meticulously try to falsify them. The term used in the history of science to refer to the particular experiment that falsifies a given theory is called the "crucial experiment," and Popper gave examples of such experiments to support his position. For instance, he noted how the Michelson–Morley experiment crucially falsified the ether theory. Because of these types of claims, Popper's position has been called "falsificationism."

According to the traditional versions of rationalism, the sources of certainty in human knowledge are either innate ideas (Descartes, Leibniz) or *a priori* conceptual frameworks (Kant). Popper, however, rejected this kind of dogmatism. In his view, there are no innate ideas or *a priori* conceptual frameworks, for all scientific theories are hypothetical – and thus fallible. So, scientific theories must bravely welcome criticism. In fact, the higher the falsifiability of a theory, the more scientific it is. Consequently, Popper's philosophy has also come to be known as "critical rationalism."

Another important figure in the school of critical rationalism, Imre Lakatos (1922–1974), was born in Hungary to a Jewish family, but moved to the UK to teach at the LSE in the wake of the Hungarian Incident of 1956. At LSE, he later succeeded Popper in the post of Professor of Logic and Scientific Method, and he also served as the editor in chief of the *British Journal for the Philosophy of Science*. Lakatos specialized in the philosophy of mathematics and general philosophy of science, and much of his work attempted to combine the ideas of Popper and Thomas Kuhn, with writings including *Proofs and Refutations* (1963–1964), *Falsification and the Methodology of Scientific Research Programmes* (1970), and *History of Science and Its Rational Reconstructions* (1971), among others. Additionally, after his death, John Worrall and Gregory Currie compiled and edited his works in a two-volume set entitled *Philosophical Papers* (1978); the first of these volumes was *Methodology of Scientific Research Programmes* and the second was *Mathematics, Science and Epistemology*.

Despite their philosophical affinities, Lakatos called Popper's philosophy of falsificationism a "naïve version of falsificationism." For Lakatos, there is no such a thing as a "crucial experiment," for scientific theories cannot be directly refuted by observations. For example, in Popper's view, the observation of the anomalous behavior of the perihelium of Mercury was the crucial experiment refuting Newtonian mechanics and supporting Einstein's theory of relativity. However, Lakatos claimed that before Einstein's theory, such observation would not falsify Newtonian celestial mechanics. Astronomers could argue instead that the anomaly was due to some undetected celestial body around Mercury that was influencing the planet's trajectory. Such an explanation could lead some scientists to attempt to search for new celestial bodies by constructing better telescopes, as was the case in the discovery of Neptune and Pluto. If still no new celestial bodies were to be found, rather than refuting Newtonian mechanics, some scientists could still argue that cosmic dust made it difficult to observe the small bodies. Therefore, they might send satellites in order to bypass the cosmic dust. Even if, finally, no celestial bodies were observed, scientists could potentially interpret the result as the effects of electromagnetic perturbations in the measurements. As the reader has probably noticed by now, this type of argumentation could, in principle, go on forever.

Thus, Lakatos argued that Popper's naïve falsificationism does not match the actual practice of scientific research. In reality, scientists are occasionally thick-skinned, and when there is conflict between theories and observations, they try to raise auxiliary hypotheses to save the theory from refutation either by experiments or observations. Sometimes, they may simply ignore an anomaly and choose to focus on other scientific issues.

Contra Popper, then, Lakatos suggested his own "sophisticated falsificationism." According to this view, the aim of the development of science is to replace old theory systems with new theory systems. Thus, only a theory, not experiments or observations, can refute another theory. Accordingly, Lakatos proposes the concept of a "scientific research programme," which is a large theoretical system with two main components: a hard core and a protective belt. On the one hand, the hard core is composed of the central concepts and fundamental laws of the theory system, for example, Newton's three laws of motion plus the law of universal gravitation in classical mechanics. On the other hand, the protective belt mainly includes auxiliary hypotheses around the hard core, for example, the number of planets in the solar system and their mass.

A research programme has the functions of both negative heuristics and positive heuristics. Negative heuristics defend the hard core from unfavorable experiments or observations through the modification or extension of auxiliary hypotheses. For example, the discovery of Neptune and Pluto successfully prevented Newtonian mechanics from being refuted by astronomical observations. Positive heuristics actively find new laws and provide explanations for new phenomena. For example, solid mechanics, fluid mechanics, and aerodynamics were developed from Newton's three laws of movement, resulting in the progress of classical mechanics.

Scientific research programmes can be either progressive or degenerating. If a research programme is able to discover new laws and deliver successful, novel predictions, then it is progressive. In contrast, if a research programme is continually challenged by anomalies but only perfunctorily modifies its protective belt, then it is a degenerating programme. In short, for Lakatos, scientific theories cannot be refuted by experiments and observations. Rather, the actual development of science takes place by replacing degenerating research programmes with progressive ones.

Historicism

As Karl Popper was to critical rationalism, so Thomas Samuel Kuhn (1922–1996) was to the next major movement in the history of philosophy of science: historicism. Born on July 18, 1922 in Cincinnati, Ohio, Kuhn went on to receive his Ph.D. in physics from Harvard University in 1949. At that time, the president of Harvard, James B. Conant, was developing the general education program and the history of science curriculum, and Kuhn taught a course in the history of science. This led to a subsequent switch for Kuhn from physics to history and philosophy of science. When he studied the rise of mechanics in the seventeenth century, Kuhn was bewildered by Aristotle's natural philosophy, realizing that in order to understand ancient physics fully, we must adopt a completely different way of thinking. Out of this realization, he raised the concept of a "paradigm."

In 1956, Kuhn moved to the University of California at Berkley, becoming Professor of the History of Science in 1961. In 1964, he joined Princeton University as Professor of Philosophy and History of Science. Finally, he moved to MIT in 1979 and retired in 1991. In the course of his long career, his research interests began with history and philosophy of science and later shifted to focus on the history of quantum mechanics, with publications including *The Copernican Revolution* (1957), *The Structure of Scientific Revolutions* (1962), *The Essential Tension* (1977), and *Black-Body Theory and the Quantum Discontinuity, 1894–1912* (1987). After his death in 1996, James F. Conant and John Haugeland edited a number of Kuhn's papers and published the collection in a volume titled *The Road Since Structure* (2000).

The central concept of Kuhn's book *The Structure of Scientific Revolutions* is "paradigm," which, as mentioned above, can be traced back to ancient Greece, as Plato used the concept to explain how the "idea" is the model of material objects. For example, the ideal circle is the paradigm of circles we actually draw. In English, a paradigm is a widely accepted model or pattern, which is frequently used in grammar to express rules of morphology and has the meaning of something imitated repeatedly. Beyond these uses, the German physicist and philosopher G. C. Lichtenberg used the concept to argue that science has a structure, and Wittgenstein also mentioned the concept and regarded the platinum prototype meter bar at the Bureau International des Poids et Mesures in Paris as the paradigm of length measurement. In his own work, Kuhn defined a scientific paradigm as a universally recognized scientific achievement that attracts many practitioners and provides them with model problems and solutions for future research. For example, in the history of science, phlogiston theory, oxygen theory, Aristotle's physics, classical mechanics, and the theory of relativity can all be seen as paradigms.

Kuhn also divided the development of science into two stages: normal science and scientific revolution. In the stage of normal science, scientists mainly do puzzle-solving research within the same paradigm. Scientific revolution, however, implies a paradigm shift: Scientists give up an old paradigm and convert to a new paradigm. In the history of science, the replacement of phlogiston theory with oxygen theory, of Aristotelian physics with Newtonian mechanics, and of classical mechanics with the theory of relativity are all cases of scientific revolution.

With that said, Kuhn's definition of paradigm is rather ambiguous, as Margaret Masterman has found at least 21 different senses of the concept (Lakatos and Musgrave 1970, p. 61). Thus, Kuhn later preferred to use the concept of a "disciplinary matrix" to capture the meaning he intended. A disciplinary matrix includes: (1) symbolic generalizations, including scientific concepts or nomenclature like the formula $F=ma$ in Newtonian mechanics; (2) shared beliefs, which include metaphysical worldviews (e.g., modern science takes the movement and interactions of particles as the cause of every phenomenon) and theoretical models (e.g., conceiving electric fluid as the flow of water); (3) shared values, as when a scientific community trains scientists to have similar standards of connoisseurship when developing and evaluating scientific theories, such as simplicity, consistency, and accuracy; and (4) exemplars, such as the demonstrative experiments typically conducted in general physics courses to illustrate basic concepts.

Kuhn's great contributions to philosophy of science were not limited to introducing historical and social factors into science, for he also proposed the thesis of incommensurability between paradigms. The term "incommensurability" can also be traced back to ancient Greece. The Pythagorean school proved that in an isosceles right triangle, there is one common measurement between the hypotenuse and the leg. Because the ratio between the hypotenuse and the leg equals the square root of 2, it cannot be written as a fractional form m/n. This kind of number violated the ancient Greeks' understanding of mathematics, so they called such numbers "irrational numbers."

Kuhn used this term to signify that different paradigms do not share sufficient common ground to make complete comparisons between them rationally. Going further, there are three different aspects of incommensurability: (1) that scientific criteria or norms are different, such as the criteria of what counts as a scientific problem; (2) conceptual change, or when the same concepts take on different meanings, like time and space; and (3) different worldviews, as some worldviews may be immutable and others radically changed (Kuhn 1970, pp. 148–150). In light of this, Kuhn argued, the choice of a given scientific theory has no rational basis, but depends on historical and social factors, which are contingent or accidental. Therefore, the objectivity and rationality of science are at stake.

The problem of relativism raised by Kuhn led in turn to various strands of thought within postmodernism, such as irrationalism, constructivism, and feminism, many of which can be seen as either developments or extensions of Kuhn's historicism. Indeed, Kuhn's historicism officially marked the end of logical positivism. In the aftermath, philosophy of science has attempted to unify historicism and logicalism and, in doing so, find a middle way.

Postmodernism

Our next major figure to introduce, Paul Feyerabend (1924–1994), was born into a middle-class Viennese family. In 1942, he was drafted into the Pioneer Corps of the German Army and was hit in his spine by a bullet in a battle with Soviet forces in 1945. In 1946, he went to Weimar to study singing and stage management and joined the Cultural Association for the Democratic Reform of Germany. In 1947, he returned to Vienna to study history and sociology, but he later changed to physics and philosophy, finally obtaining his Ph.D. in philosophy in 1951. In the same year, he was granted a British Council scholarship to study under Wittgenstein at Cambridge University, but due to Wittgenstein's death, he instead moved to the LSE, where he studied quantum physics and Wittgenstein's thought under Popper's supervision. Through this experience, he was influenced by Popper's critical rationalism and later criticized logical positivism for a portion of his career. In 1959, he emigrated to the United States and became a professor at the University of California, Berkley, where he experienced a gradual shift in his views. After 1970, he embraced Kuhn's historicism and argued against Popper's falsificationism. He died on February 11, 1994, his notable works by that point including *Against Method* (1975), *Science in a Free Society* (1978), *Farewell to Reason* (1987), and *Three Dialogues on Knowledge* (1991).

Feyerabend emphasized the incommensurability thesis and extended the irrational implications of Kuhn's historicism to the extreme, in the process becoming a radical irrationalist. He famously argued against scientific methods, instead promoting epistemological anarchism and offering the slogan "Anything goes." In *Against Method*, he writes: "anarchism, while perhaps not the most attractive political philosophy, is certainly excellent medicine for epistemology, and for the philosophy of science." He then goes on to expound two reasons for such a claim:

> The first reason is that the world which we want to explore is a largely unknown entity. We must, therefore, keep our options open and we must not restrict ourselves in advance... The second reason is that a scientific education as described above...cannot be reconciled with a humanitarian attitude. It is in conflict 'with the cultivation of individuality which alone produces, or can produce, well-developed human beings.'
>
> (Feyerabend 1975, pp. 17–20)

Though Feyerabend's philosophy was quite radical, nevertheless, as a natural development of historicism, it raised and continues to inform a range of pressing questions that we must take seriously. His return to the historical materials, demand for deep research into the external history of science including social, political, and economic factors, and emphasis on the historicity, specificity, and limit of scientific methods are all still of great significance in philosophy of science.

The new wave

The final historical trend in the development of the history of philosophy of science, which brings us to the present day, is the so-called "new wave," with representative figures being Larry Laudan, Hilary Putnam, and Dudley Shapere. Larry Laudan was born on October 16, 1941 in Austin, Texas, and he received a bachelor's degree in physics from the University of Kansas in 1962 and a Ph.D. in philosophy from Princeton University in 1965. Throughout his career, he has taught at a range of institutions, including University College London, the University of Pittsburgh, the Virginia Polytechnic Institute and State University, and the University of Hawaii, though he has mainly lived in Mexico since the year 2000.

Laudan is in an inheritor of the tradition of historicism, but he approaches the rationality and progress of science from the positions of empiricism and pragmatism. For Laudan, the aim of science is not to seek truth but to solve problems. As he claims, the new paradigm of science has more puzzle-solving power than older paradigms had, so the development of science can be seen as rational and unceasingly progressive.

Hilary Putnam (1926–2016) studied logic at Harvard University and then philosophy under Reichenbach at UCLA, where he obtained his Ph.D. in 1951. He taught variously at Northwestern University, Princeton University, and MIT, finally moving to Harvard University in 1965, and he was a Fellow of the American Academy of Arts and Sciences and a Corresponding Fellow of the British Academy. His writings include *Mathematics, Matter, and Method* (1975), *Mind, Language, and Reality* (1975), *Reason, Truth, and History* (1981), *Realism and Reason* (1983), and *The Many Faces of Realism* (1987). Putnam transformed the traditional theory of meaning and, following Saul Kripke, proposed the causal chain theory of reference, thus defending the rationality and reality of science by insisting on convergent realism (Putnam 1982, pp. 195–200). Convergent realism suggests the following: (1) Scientific terms in a mature science genuinely refer; (2) theories of a mature science are approximately true; (3) the same concept in different scientific theories can have the same reference; and (4) the above propositions are not necessarily true, but are a scientific interpretation of the relation between scientific theories and research objects (Shu and Qiu 2007, pp. 270–287).

Finally, Dudley Shapere (1928–2020) was born on May 27, 1928 in Harlingen, Texas, and he obtained his Ph.D. in 1957 from Harvard University. He taught consecutively at the Ohio State University, the University of Chicago, the University of Illinois, the University of Maryland, and Wake Forest University, and he also served on the editorial board for *Philosophy of Science*. His main writings include *Philosophical Problems of Natural Science* (1965), *Galileo: A Philosophical Study* (1974), and *Reason and the Search for Knowledge* (1984).

Shapere raised the criticism that logical positivism and historicism fall into two opposite, but equally wrong, philosophical tendencies: absolutism and relativism. Absolutism and presuppositionalism hold one assumption in common: "there is something which is presupposed by the knowledge-acquiring enterprise, but which is itself immune to revision or rejection in the light of any new knowledge or beliefs acquired" (Shapere 1984, p. 205). He further mentioned four sorts of presuppositionalism: (1) ontological, which claims that some ontological principles must be accepted before scientific inquiry (e.g., the law of uniformity of nature, the principle of simplicity, etc.); (2) methodological, which holds that the "the scientific method" is the means by which scientific knowledge is obtained; (3) logical, which applies logic in all scientific reasoning; and (4) conceptual, which insists that some unchanged concepts are employed in science. On the other hand, because logical positivists seek metascientific language and inference rules for scientific knowledge, they are met with the problem of absolutism (Shapere 1984, pp. 205–260).

According to Shapere, historicism, when rejecting the absolutism of logical positivism, successfully reveals the historicity and limitations of human knowledge, but it denies the objectivity and progressiveness of scientific knowledge, which leads to relativism. To reject absolutism and avoid relativism, Shapere thus proposed the concept of "domain of inquiry," which is a unified body of information that still embraces scientific realism (Jiang 1984, pp. 220–259).

Indeed, since the 1980s, scientific realism has been an emerging philosophical issue, concerned with whether or not the theoretical entities in modern science actually exist. Scientific realists include Wilfrid Sellars, John Searle, Grover Maxwell, and Ian Hacking, while anti-realists include Bas van Fraassen (empirical constructivism) and Arthur Fine (Natural Ontological Attitude). For more details on this issue, see Chapter 9.

Notes

1 Some Chinese philosophers trace philosophy of science back to the debate between rationalism and empiricism, or even ancient Greek (Xia and Shen 1987). But to emphasize philosophy of science in a modern context, this book will focus on philosophy of science since the twentieth century.

2 The different term "logicism" has been used to refer to the philosophical school that regards logic as the foundation of mathematics (Tennant 2017).
3 According to Waismann, however, Schlick disagreed with positivism on three different counts, so Schlick himself preferred the term "logical empiricism" (recounted in Hong 1999, 122–123).

3 Introduction to logic

Logic is the most basic training in Western philosophy of science, and logical analysis is one of the most fundamental methods in this philosophical realm of research. Accordingly, this chapter will introduce some basic knowledge of logic, which will give the reader some logical background for better understanding the discussions throughout the rest of the book.

Logic and argument

In daily life, we make many inferences. For example, from the statements "All men are mortal" and "Socrates is a man," we infer that "Socrates is mortal," or from the statement "All observed swans are white," we infer that "All swans are white." If an inference is articulated with language, we call it an argument.

Arguments are composed of premises and conclusions. For example, in the argument "All men are mortal; Socrates is a man; therefore, Socrates is mortal," "All men are mortal" and "Socrates is a man" are premises, while "therefore, Socrates is mortal" is the conclusion. An argument is constructed to draw a conclusion from its inference.

Logic studies the support for linking a premise to a conclusion in an argument. Among the forms, deduction is inference with necessity, while induction is inference without necessity. These two types of inferences can be called "logic" in a broad sense. However, the development of deductive logic has been more rapid, and the research relative to this type has gone into significantly greater depth, so in a narrow sense "logic" is often used in reference to deductive logic more specifically. Likewise, logic as it is discussed in this chapter refers more specifically to deductive logic.

In deductive logic we focus on the relation between premises and conclusions, regardless of their truth or falsity. For example, in the inference "All members of Tsinghua University are crazy; Wang Wei is a member of Tsinghua University; therefore, Wang Wei is crazy," both the premise "All members of Tsinghua University are crazy" and (hopefully) the conclusion "Wang Wei is crazy" are false, but the argument itself is still *logically* correct. By contrast, all the phrases in the argument "Philosophers of science are philosophers; C. G. Hempel is a philosopher; therefore, C. G. Hempel is a philosopher of science" are true, but the logical form of the argument is wrong. Logic studies only the inferential mode of argumentation. If a conclusion can be necessarily inferred from the premises, then the argument is logically "valid."[1]

Starting in the twentieth century, the study of logic developed more and more rapidly, and many systems of logic have subsequently been developed. Among them, propositional logic and quantificational logic, also called predicate logic, constitute the standard systems of logic. In addition, there are also modal logic, many-value logic, deontic logic, and others. Here, we only introduce the basics of first-order predicate logic.

Propositional logic

Propositional logic deals with the truth values of compound propositions and the validity of inference between propositions, which correspond to both the propositions and the inferential relations between them. The word "proposition" was itself proposed by the logician Frege as a designation for the meaning of a sentence. Thus, the difference between a sentence and a proposition is that the former is a part of language, whereas the latter entails the meaning expressed by the sentence. For example, "Snow is white" and "La neige est blanche" are different sentences linguistically, but they express the same proposition.

Usually, we can directly find the truth value of an elementary proposition, for example, "It was sunny in Beijing on February 14, 2003" or "Sixty students took the course 'Modern Western Philosophy of Science.'" However, sometimes we wish to express more complex meanings, such as "If you don't behave well, I will not take you out to play" or "Today is sunny, and I feel good." These kinds of propositions are called "compound propositions," which are made up of simple propositions and logical connectives. Their truth values depend on the truth value of the elementary propositions and the logical connectives, and can be determined by logical calculation.

The main logical connectives in propositional logic include: "not," designating negation, which is expressed as $\neg$ (also written as $\sim$ in some logic textbooks); "and," designating conjunction and expressed as $\wedge$ (also as $\cdot$); "or," designating disjunction and expressed as $\vee$; "if...then...," designating implication and expressed as $\rightarrow$ (also as $\supset$); and "if and only if..." (usually written as "iff"), designating equivalence and expressed as $\leftrightarrow$ (also as $\equiv$). The logical relations expressed by these logical connectives can be represented by a truth table. The values 1 and 0 in the table represent "true" (T) and "false" (F), respectively.

Furthermore, the connective "negation" denotes the negative relation to a proposition. For instance, the negation of the proposition "Tomorrow it will rain" is "Tomorrow it will not rain." When a proposition is true, its negation will be false, and vice versa. With all this in mind, the truth table of the logical negation is:

Table 3.1 Truth table for negation

p	$\neg p$
1	0
0	1

The connective "conjunction" signifies that when two propositions are true, the composite proposition is true; in any other case, the composite proposition is false. For example, the composite proposition "Tomorrow it will rain and the temperature will be less than 15°C" will be true only in the event that both propositions "Tomorrow it will rain" and "Tomorrow the temperature will be less than 15°C" are true; otherwise, it will be false. Thus, the truth table for the conjunction is:

Table 3.2 Truth table for conjunction

p	q	$p \wedge q$
1	1	1
1	0	0
0	1	0
0	0	0

It should be noted that a conjunction connects two propositions rather than phrases. For example, "Moritz Schlick and Rudolf Carnap are members of the Vienna Circle" needs to be translated into "Moritz Schlick is a member of the Vienna Circle, and Rudolf Carnap is a member of the Vienna Circle." In addition, terms expressing the turning relationships in ordinary language, such as "but," can also be regarded logically as conjunctions. For example, "It will rain tomorrow, but the temperature will be very high" and "It will rain tomorrow, and the temperature will be very high" are logically equivalent.

Disjunction signifies that when two propositions are false, the resulting composite proposition is also false; in any other case, the composite proposition is true. For instance, "Tomorrow it will rain or the temperature will be less than 15°C" will be false only in case that both propositions are false; otherwise, the composite proposition will be true. The truth table for disjunction in this way is:

Table 3.3 Truth table for disjunction

p	q	$p \vee q$
1	1	1
1	0	1
0	1	1
0	0	0

There are also differences between "or" in logic and in ordinary language. In some scenes of daily life, we use "or" to indicate that only one of them can be selected. For example, after dinner, the waiter asks, "Tea or coffee?"; this usually implies that we are to choose only one item. If we persist with a strict logical meaning of disjunction, it would not be so polite for us to choose both.

Logical implication can be written as "if…then…," so it can be divided into antecedent and consequent. When the antecedent is true and the consequent is false, the whole proposition is false; otherwise, it is true. For example, the proposition "If you are admitted to Harvard University, then I will treat you to dinner" is false only if you are admitted by Harvard and I do not invite you to have dinner. In any other case, the proposition is true. The truth table of logical implication is as follows:

Table 3.4 Truth table for implication

p	q	$p \rightarrow q$
1	1	1
1	0	0
0	1	1
0	0	1

Logical implication is quite different from daily language. According to the truth table of implication, all cases are true that do not include both the antecedent as true and the consequent as false. Thus, the propositions "If 1+1=3, then Peking University is in Beijing (or in Nanking)" and "If the Earth is round (or cubic), then 1+1=2" are both true. Because in the first case the antecedent is false, and in the second case the consequent is true, the compound propositions should be true according to the truth table. Obviously, this goes against the use of ordinary language. Therefore, some logicians suggest that implication $A \rightarrow B$ should be understood as $\neg(A \wedge \neg B)$ or $\neg A \vee B$.

A double implication, also known as a biconditional statement, can be read as "if and only if" to indicate that the compound proposition is true when two propositions are equally true or false, but is false under any other conditions. The according truth table of equivalence relation is as follows:

Table 3.5 Truth table for double implication

p	q	$p \leftrightarrow q$
1	1	1
1	0	0
0	1	0
0	0	1

The above five logical conjunctions can be defined mutually. For examination, A→B can be defined as ¬A∨B, and A∨B can be defined as ¬A→B. So, in principle, we can define all logical connectives in just two ways, for example, ¬ and ∨, or ¬ and →.

In the truth table, if a logical expression is true in any case, we call it a tautology; if it is false in any case, we call it a contradiction. For example, p∨¬p is a tautology, whereas p∧¬p is a contradiction. Their truth table is:

Table 3.6 Truth table for p∨¬p and p∧¬p

p	$p \lor \neg p$	$p \land \neg p$
1	1	0
0	1	0

Truth tables can help us not only to calculate the truth values of compound propositions, but also to judge the validity of some inferences. According to the definition of validity, the conclusion must be true if all the premises are true. We can take the implication between the premises and the conclusion. If the whole expression is a tautology, then in all cases the conclusion is true when the premises are true, so the inference is valid. If not, in some cases the conclusion may be false when the premises are true, rendering the inference invalid.

For example, the father says to the child, "If you behave well, then I will take you to the cinema." When the child behaves well, can we logically deduce that "The father will take the child to the cinema"? Or if the child does not behave well, can we logically conclude that "The father will not take the child to the cinema"? We can translate these two inferences into logical expressions: p→q, p, ∴q and p→q, ¬p, ∴¬q, which can be written as logical implications in the form ((p→q)∧p)→q) and ((p→q)∧¬p)→¬q), respectively. Since the former is a tautology, the inference is valid; as the latter is not a tautology, the corresponding inference is thus not valid. The former is called the inference of "affirming the antecedent" (*modus ponens*), and its sibling valid inference is "denying the consequent" (*modus tollens*), whose logical form is: p→q, ¬q, ∴¬p). The latter is called the fallacy of "affirming the consequent," and its sibling fallacy is "denying the antecedent" (the logical form is: p→q, ¬p, ∴¬q). Their truth table is as follows:

Table 3.7 Truth table for four inferences of logical implication

p	q	*Affirming the antecedent* $((p \to q) \land p) \to q$	*Denying the consequent* $((p \to q) \land \neg q) \to \neg p$	*Denying the antecedent* $((p \to q) \land \neg p) \to \neg q$	*Affirming the consequent* $((p \to q) \land q) \to p$
1	1	1	1	1	1
1	0	1	1	1	1
0	1	1	1	1	0
0	0	1	1	0	1

The truth table calculation of propositional logic is a mechanical procedure that can be carried out by computer. The formation rules of propositional logic can also be represented by mechanical procedures, so that a computer can determine whether or not a formula is well formed in accordance with the logical grammar:

(1) p, q, r, etc. are well-formed formulas.
(2) If A and B are well-formed formulas, then ¬A, (A∧B), (A∨B), (A→B), and (A↔B) are well-formed formulas.
(3) Any other expression is not a well-formed formula.

Syllogism

Propositional logic can only deal with the logical relations between propositions, not their internal structure. For instance, "All Guangzhou citizens are Cantonese; all Cantonese are Chinese; therefore, all Guangzhou citizens are Chinese" is a valid inference. However, if we express it by propositional logic in the form "p∧q, ∴r," it is not a tautology.

In the past, syllogism could deal with such inferences. Syllogism was founded by the ancient Greek philosopher Aristotle and later became the chief model of traditional logic. Syllogism deals with four sentence patterns:

A sentence (universal affirmative): All F are G.
E sentence (universal negative): No F is G.
I sentence (particular affirmative): Some F are G.
O sentence (particular negative): Some F are not G.

Syllogism consists of three terms and three sentences (two as premises, one as conclusion, whose sentence pattern is one of the above four). The subject in the conclusion is called "subject" (represented as S) and the predicate is called "predicate" (P). They may also be called "end terms." The term shared by the premises is called the "middle term" (M). For example, the syllogism "All Guangzhou citizens are Chinese" has the following structure:

All Guangzhou citizens (S) are Cantonese (M).	A sentence.
All Cantonese (M) are Chinese (P).	A sentence.
All Guangzhou citizens (S) are Chinese (P).	A sentence.

Since every sentence in a syllogism may have four possible patterns, and the subject and predicate in the premise can be interchanged, there can thus be 256 possible variations in total. The validity of a syllogism can be judged by two methods: the rule method and the Venn diagram method.

The rule method generalizes the 256 possible variations and holds that the syllogism is valid only if it satisfies the following three rules:

(1) The middle term is distributed once.
(2) No end term is distributed once.
(3) The number of negative premises is equal to the number of negative conclusions.

A term is "distributed" when we can have some idea of its complete extension. For example, from "All F are G," we can know F's extension belongs to G, so F is distributed. From another example, "No F is G," we can know F's extension does not belong to G and vice versa, so both F and G are distributed. In the case that "Not every F is G," we can know G's extension does not include the particular F that is not G, so G is distributed. In the four sentence patterns, the other words are not circumscribed. All the other terms in the four sentence patterns are not distributed. The relationship between sentence pattern and "distributed" is as follows (distributed is represented by "d," while "u" means "not distributed"):

A sentence: All A_d are B_u.
E sentence: No A_d is B_d.
I sentence: Some A_u is B_u.
O sentence: Some A_u is not $B_{d.}$

According to the rule method, we can judge the validity of all syllogisms. Taking the example "All Guangzhou citizens are Cantonese; all Cantonese are Chinese; therefore, all Guangzhou citizens are Chinese," the middle term "Cantonese" is only distributed once, so Rule (1) is satisfied. The end terms "Guangzhou citizens" and "Chinese" are distributed twice and none, so Rule (2) is satisfied. Furthermore, the argument has no negative premise and no negative conclusion, so Rule (3) is satisfied. In this way, the argument satisfies all three rules, so it is valid.

In contrast, the syllogism "All Guangzhou citizens are Cantonese; some Guangzhou citizens are very rich; therefore, some Guangzhou citizens are very rich" violates the first rule because the middle term ("Cantonese") is not distributed at all. In consequence, we can directly judge that the syllogism is not valid. After all, there is a logical possibility that all rich Cantonese happen not to be in Guangzhou.

The Venn diagram method is used to study the validity of syllogisms by drawing pictures. Thus, in the diagram, the subject term, predicate term, and middle term are represented by three different circles and the oblique lines represent the empty region. The four sentence patterns can be represented by the corresponding diagrams:

All syllogisms can be expressed with Venn diagrams. If the graph of conclusion is already contained in the graph of the premises, then the argument is valid. If the graph of the premises cannot contain the graph of conclusion, then it is not valid (Salmon 1984).

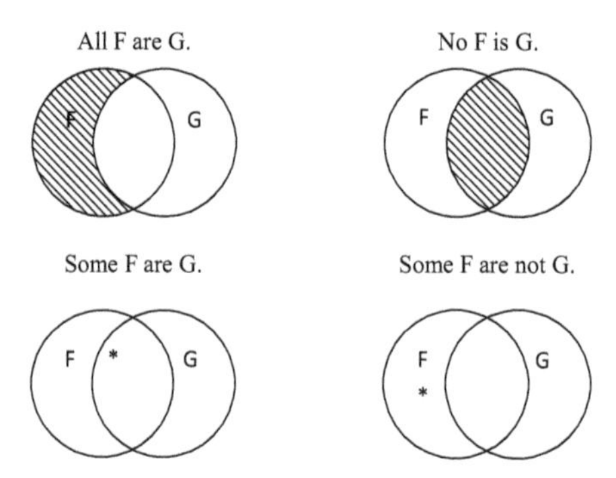

Figure 3.1 Venn diagrams of four sentence patterns

Quantificational logic

Propositional logic is unable to express the internal logical relation within propositions, so we need quantificational logic for this task. Quantificational logic mainly introduces two quantifiers: the universal quantifier "All" and the existential quantifier "Exist" to express the logical relation within propositions.

The universal quantifier "All" is written in logical form as $\forall x$. The symbol $\forall$ is the inversion of the first letter of "All." In some books, the universal quantifier is written as (x). This gives us the phrase "All x are…" The existential quantifier "Exist," also written as "for some" in some textbooks, is represented in logical form as $\exists x$. This symbol $\exists$ is a back-to-front version of the first letter of "Exist," and it means "There exists x…" (or "For some x…").

The vocabulary of quantificational logic includes five parts: (1) propositions, represented by p, q, r, etc.; (2) predicate, represented by F, G, H, etc.; (3) constant, represented by a, b, c, etc.; (4) variable, represented by x, y, z, etc.; and (5) quantifier, represented by $\forall x$, $\exists x$ (x can be replaced by other variables). In this way, all syllogisms can be expressed by quantificational logic. For example, the four sentence patterns of syllogisms can be written in such quantificational logical forms:

A sentence: All F are G.	$\forall x(Fx \rightarrow Gx)$
E sentence: No F is G.	$\forall x(Fx \rightarrow \neg Gx)$
I sentence: Some F are G.	$\exists x(Fx \wedge Gx)$
O sentence: Some F are not G.	$\exists x(Fx \wedge \neg Gx)$

Therefore, the validity judgment of syllogisms can be written as reasoning in the form of quantificational logic, and thus be examined. The examining methods mainly include natural deduction and the method of axiomatization.

Axiomatic system of first-order logic

Propositional logic and the first-order quantificational logic can be reformulated with an axiomatic system. There are eight axioms and two inference rules in total. The axioms are as follows:

$\Phi 1$: $\alpha\rightarrow(\beta\rightarrow\alpha)$
$\Phi 2$: $(\alpha\rightarrow(\beta\rightarrow\gamma))\rightarrow((\alpha\rightarrow\beta)\rightarrow(\alpha\rightarrow\gamma))$
$\Phi 3$: $(\sim\alpha\rightarrow\beta)((\sim\alpha\rightarrow\sim\beta)\rightarrow\alpha)$
$\Phi 4$: $\forall x\alpha\rightarrow\alpha(t/x)$
$\Phi 5$: $\alpha\rightarrow\forall x\alpha$
$\Phi 6$: $t\equiv t$
$\Phi 7$: $(t_1\equiv S_1\rightarrow\ldots\rightarrow t_n\equiv Sn)\rightarrow f(t_1,\ldots,t_n)\equiv f(S_1,\ldots,Sn)$
$\Phi 8$: $(t_1\equiv S_1\rightarrow\ldots\rightarrow t_n\equiv Sn)\rightarrow P(t_1,\ldots,t_n)\equiv P(S_1,\ldots,Sn)$

The inference rules are as follows:

MP: $\alpha\rightarrow\beta$, $\alpha\vdash\beta$
RD: $\vdash\alpha$, $\vdash\forall x\alpha$

In the twentieth century, the logician Kurt Gödel proved the completeness of first-order logic – that is, that all the true sentences in first-order logic can be derived from the above axioms and inference rules. Here we need only give the proof of A→A as an example:

(1) $\vdash(A\rightarrow((A\rightarrow A)\rightarrow A))\rightarrow(((A\rightarrow(A\rightarrow A))\rightarrow(A\rightarrow A))$	Axiom 2
(2) $\vdash A\rightarrow((A\rightarrow A)\rightarrow A)$	Axiom 1
(3) $\vdash(A\rightarrow(A\rightarrow A))\rightarrow(A\rightarrow A)$	(1)(2)MP
(4) $\vdash A\rightarrow(A\rightarrow A)$	Axiom 1
(5) $\vdash A\rightarrow A$	(3)(4)MP

Logic and language analysis

The reason philosophy of science attaches so much importance to logic – after all, logical positivism and logical empiricism are even named "logical" – is that logic can be widely used to analyze ordinary language and clarify linguistic ambiguities. For example, there was a famous debate in ancient China called the "White Horse Dialogue," which centered around the phrase "White horses are not horses" (白马非马). According to custom regulations at the time, horses were not allowed to exit a gate. But the ancient philosopher Gongsun Long (公孙龙) argued that white horses are not horses, for white horses must be white, but horses can be of other colors; as a result, his white horse should be allowed to exit. The argument "A white horse is not a horse" thus uses the ambiguity of ordinary language.

In fact, we readily recognize that there are many usages of "to be" in ordinary language. Even the simple statement "A is B" can be written in at least five logical forms:

(1) $a \in B$, i.e., an element belongs to a set, e.g., "Wang Wei is a professor."
(2) $A \subset B$, i.e., a set is included in another set, e.g., "Philosophers of science are philosophers."
(3) $a=b$, i.e., two objects have the same identity, e.g., "The morning star is the evening star."
(4) $A=B$, i.e., the two sets have the same extension, e.g., "The animals with hearts are the animals with kidneys."
(5) Ba, i.e., an object has some property, e.g., "Snow is white."

Therefore, "to be" has at least five logical forms, yet these cannot be distinguished properly and precisely in ordinary language. The sophistry of "A white horse is not a horse" confuses the second and fourth usages. When Gongsun Long says that white horses and horses are not identical, he uses "to be" in the fourth sense, but when he argues that his white horse is not a horse, he uses "to be" in the second sense. If we have regular training in the use of logic, this kind of language trick is easy to expose.

Note

1 A sound argument requires not only that the logical form is valid, but also that all the premises are true.

4 Criteria of cognitive significance

At present, Western philosophy can be roughly divided into two camps: Anglo-American analytical philosophy and continental speculative philosophy. Anglo-American analytical philosophy can be traced back to British philosophers like John Locke, George Berkeley, and David Hume. As empiricists *par excellence* who emphasize that all knowledge starts from daily experience (and thus oppose armchair speculation), their style of writing strives for clarity and understandability. Because philosophy of science mainly inherits the traditions of empiricism and positivism, it basically belongs to the scope of Anglo-American analytical philosophy.

The main figures of continental speculative philosophy are G. W. F. Hegel, Martin Heidegger, Edmund Husserl, Henri Bergson, and so on. They have had a wide and deep influence in continental European countries, especially Germany and France. Most of this persuasion tries to establish a theoretical system that provides a general worldview for everything. Hegel is particularly prominent in this respect, as he attempted to use dialectics to establish an all-inclusive system for explaining all phenomena, such as nature, history, art, religion, etc. During the nineteenth century, Hegel's philosophy prevailed throughout much of Europe.

A. J. Ayer has claimed that philosophy in the twentieth century started as a "reaction to Hegel" (Ayer 1982). Logical empiricists especially began to question the meaning of language, essentially refuting the value of speculative philosophy altogether. For example, Reichenbach quotes a passage from Hegel's *Philosophy of History*:

> Reason…is *Substance*, as well as *Infinite Power*; its own *Infinite Material* underlying all the natural and spiritual life which it originates, as also the Infinite Form, — that which sets this Material in motion… . Reason is the *substance* of the Universe; viz. that by which and in which all reality has its being and subsistence.
>
> (Hegel 1956, p. 9)

Then, assessing such Hegelian speculation, Reichenbach concludes that this passage is neither right nor wrong, but it is nonetheless vague and meaningless.

But why is speculative philosophy meaningless? What kind of sentence, then, would be considered meaningful? This brings us to the empiricist criteria of meaning.

Proposing the criteria of meaning

As early as the eighteenth century, the philosopher Hume was clearly opposed to meaningless speculation, even if he did not clearly define what should be the standard criteria of meaning in response. He wrote:

> When we run over libraries, persuaded of these principles, what havoc must we make? If we take in our hand any volume; of divinity or school metaphysics, for instance; let us ask, Does it contain any abstract reasoning concerning quantity or number? No. Does it contain any experimental reasoning concerning matter of fact and existence? No. Commit it then to the flames: For it can contain nothing but sophistry and illusion.
>
> (Hume 2007b, p. 120)

Running down the centuries, the founder of positivism, Auguste Comte, took more or less the same view. Then we come to logical positivism, the inheritor of both the empiricist and the positivist traditions, with such representatives as Moritz Schlick, who claims:

> The great contemporary turning point is characterized by the fact that we see in philosophy not a system of cognitions, but a system of acts; philosophy is that activity through which the meaning of statements is revealed or determined. By means of philosophy statements are explained, by means of science they are verified.
>
> (Schlick 1959, pp. 53–59)

Rudolf Carnap also took the position that the logical analysis of language will show that all the assertions of metaphysics are meaningless, thus completely eliminating metaphysics (Carnap 1932, pp. 60–81). He argued, for instance, that if we want to judge whether or not a word is meaningful, we should first clarify the syntax of the word – that is, how it appears in the elemental sentence form. For example, the elemental sentence form of "stone" is "X is a stone." Second, he maintained, for an elemental sentence containing the word, a question must be answered, which can be expressed in different ways (Carnap 1932, p. 62):

(1) From what sentences is S *deducible*, and what sentences are deducible from S?
(2) Under what conditions is S supposed to be true and under what conditions false?
(3) How is S to be *verified*?
(4) What is the *meaning* of S?

Continuing with the example, the elemental sentence form of "stone," "X is a stone," can be directly observed to be true or false, so it is meaningful. Likewise, though a word like "arthropod" is more abstract and complex, its elemental sentence form "X is an arthropod" can be deduced from the sentences "X is an animal," "X has a segmented body," "X has jointed legs," and on down the list of qualities. Since all these sentences are "observational sentences" or "record sentences," "arthropod" is also a meaningful word.

Carnap summarized the results of the above analysis as follows: Let "a" be any word, and "S(a)" be the elementary sentence in which the word occurs. The necessary and sufficient conditions for "a" to be meaningful are the following expressions, which actually say the same thing (Carnap 1932, pp. 64–65):

(1) The *empirical criteria for* "a" are known.
(2) It has been stipulated from what protocol sentences "S(a)" is *deducible*.
(3) The *truth conditions for* "S(a)" are fixed.
(4) The method of *verification* of "S(a)" is already known.

Furthermore, he would contend, if we want to invent our own new words, we should also meet the above requirements.

With all that said, much of the terminology used in metaphysics cannot meet the above conditions. By way of example, Carnap took the word "principle." "Principle" derives from the Latin *principium*, originally meaning "temporal precedence," but metaphysicians expanded its original meaning and provided an additional meaning of "metaphysical precedence." But what is "metaphysical precedence"? Metaphysicians have no clear standard for defining this, for when it comes to clarifying the conditions under which the proposition "X is the principle of the world" is true or false, metaphysicians often have vague answers. For this reason, Carnap concluded that the word "principle" as it is understood in metaphysics is meaningless.

After clarifying the meaning of words, Carnap further proposed the criteria of meaning for sentences. First of all, sentences containing any meaningless words are meaningless. If "principle" is a meaningless word, then "Water is the principle of the world" is a meaningless sentence. Furthermore, even if all the words in a sentence are meaningful, the whole sentence may be meaningless, such as if it violates the rules of grammar. Carnap gives the example of "Caesar is and." Because the word "and" is syntactically a conjunction and cannot be used as a predicate, the sentence is meaningless. Another example is "Caesar is a prime number." Although it accords with proper English syntax, "prime number" is a feature describing numbers and cannot be matched with a person's name, so the whole sentence is still a meaningless "pseudo-statement."

Of course, the above example of a pseudo-statement is quite obvious. In Carnap's view, however, sophisticated metaphysics is also guilty of similar pseudo-statements, only disguised more cleverly and delicately. To illustrate this, Carnap quoted the following passage from Heidegger's essay "What is Metaphysics?":

> What is to be investigated is *being* only and – nothing else; being alone and further – *nothing*; sole being, beyond being – *nothing*; *What about this Nothing*? … *Does the Nothing exist only because the Not, i.e. the Negation, exists*? Or is it the other way around? *Does Negation and the Not exist only because the Nothing exists*? … We assert: *the Nothing is prior to the Not and the Negation*.... Where do we seek the Nothing? How do we find the Nothing.... We know the Nothing.... *Anxiety reveals the Nothing*.... That for which and because which we were anxious, was "really" – nothing. Indeed: the Nothing itself – as such – was present.... *What about this Nothing*? – *The Nothing itself nothings*.
>
> (Cited in Carnap 1932, p. 69)

The paragraph seems profound and thoughtful; nevertheless, in Carnap's thinking, it is meaningless. This is because logic stipulates that the logical form of "Not" is $\neg$ and "Exist" is $\exists x$. When we say "Unicorns do not exist" or "Aliens do not exist," the statements can be written in the logical form of $\neg\exists x\ f(x)$. Still, it is difficult for us to express such a notion as "Nothing" in a logical form. Especially something as enigmatic as "The Nothing itself nothings" cannot be written as a well-formed formula in accordance with first-order logic.

Following ideas such as these, logical empiricism, along with logical atomism and logical positivism, emphasizes that philosophy is essentially language analysis. For those subscribing to such a view, meaningful sentences include two parts: analytic propositions, whose truth and falsehood can be determined only by logic and grammar, and which include analytic truth and contradiction sentences; and synthetic propositions, which describe the world and have empirical content. In this way, we might say, the work of philosophers becomes similar to that of postmen. They deliver the analytic propositions to mathematicians, logicians, and linguists, and the synthetic propositions are delivered to empirical scientists; all other sentences, including those of a metaphysical nature, are meaningless. As for the meaningless metaphysics, earlier philosophers such as Hume and Comte took a radical attitude and suggested burning such books (or letters, following our postal analogy). By contrast, figures like Schlick have taken a more moderate position, regarding metaphysics instead as a kind of poetry, which expresses human attitudes towards life and thus still has some significance in our lives. For example, the famous Tang poem "To thirty thousand feet, My white hair would grow,/'Cause like this long is my woe" (白发三千丈，缘愁似个长) is not a description of the world and is therefore neither true nor false in value. Nevertheless, it may still bring consolation to our lives through its poetic expression. Similarly, Schlick has held that Plato's *Republic* and Augustine's *Confessions* provide very useful life lessons. So even though they do not satisfy the logical empiricist criteria of meaning, those ancient philosophical classics still have great significance.

Carnap also regards metaphysics as "a substitute, albeit an inadequate one, for art." He claims that "Metaphysicians are musicians without musical ability" and expresses appreciation for Nietzsche's use of poetry to write philosophy. For this reason, later logical positivists define the word "meaning," as it is used in the phrase "criteria of meaning," as "cognitive meaning," and thus "significance" as "cognitive significance."[1] Metaphysics has no cognitive significance, but it does have significance for life.

In short, logical positivists propose criteria of cognitive significance to exclude metaphysics from philosophy. But how to reject meaningless statements? For this, it is necessary to put forward the criteria of cognitive significance.

Testability criteria

To begin, the primary criterion of cognitive significance is testability: A statement has cognitive significance if and only if it can be tested. Naturally, due to the different ways of testing, the testability principle can be divided into the verification principle, the falsifiability criterion, and the confirmability criterion.

The verification principle

Logical positivists first propose the verification principle. As Moritz Schlick puts it, "The meaning of a proposition is its method of verification" (Schlick 1936, pp. 339–369).

The logical form of the verification principle can be written as follows, where O stands for the observational proposition; O_1, O_2 … O_n are a finite number of non-contradictory observational propositions, usually required to be mutually independent; and S is the statement to be investigated:

$$\{O_1, O_2, \ldots O_n\} \vdash S$$

The verification principle argues that a statement has cognitive significance if and only if it can be verified by a series of observational propositions. For example, "Beijing was very cold in February 2003" can be broken down into "Beijing was very cold on February 1, 2003," "Beijing was very cold on February 2, 2003," and so on, until "Beijing was very cold on February 28, 2003." All of these can be verified by direct observation, so the sentence has cognitive significance.

Although the verification principle is rather intuitive, it nonetheless has a fatal omission: It ignores strict universal propositions. Universal propositions can usually be written in the form "All F are G." When the number of

F is infinite, we call it a "strict universal proposition." Obviously, no matter how many samples we observe, they are finite in number, hence strict universal propositions cannot be verified logically. For example, with the proposition "All metals conduct electricity," because the number of metals may be infinite, no matter how many metals we examine, we cannot verify the proposition.

Scientific laws usually appear in the form of strict universal propositions, so few can be verified. According to the verification principle, therefore, these scientific laws are devoid of cognitive significance. Undoubtedly, such a conclusion would be hard for us to accept.

In addition, another kind of counterexample can be constructed. Consider any proposition S deducible from a sequence of observational statements – e.g., "There are eight planets in the solar system" – and a meaningless statement N – e.g., "Caesar is a prime number." From this, we can get the disjunction statement S∨N – that is, "There are eight planets in the solar system, or Caesar is a prime number."

Since S is deducible from a sequence of observational statements, according to the logical rules, so is S∨N. Consequently, S∨N satisfies the verification principle and thus has cognitive significance. But intuitively we would think that if a compound sentence contains any meaningless sentence, the whole sentence should be meaningless. So, this counterexample reminds us that we can make any meaningless statement satisfy the verification principle and become part of a statement with cognitive significance by means of a disjunctive expression.

Opponents of verificationism also propose that the negation of a true proposition must be false, and vice versa. For this reason, we can also affirm that if a proposition has cognitive significance, so does its negation. But according to the verification principle, existential propositions – e.g., "Some metals conduct electricity" – can be verified and are therefore meaningful. However, the negation of many existential propositions – e.g., "All metals do not conduct electricity" – cannot be verified because they are strict universal propositions and thus have no cognitive significance. The negation of a meaningful statement is meaningless, as this obviously violates our understanding of truth tables in general logic. Therefore, although the verification principle was the first criterion of cognitive meaning, it is quite problematic itself.

The falsifiability criterion

Popper has been commonly regarded as the advocator of the next criterion, that of falsifiability, although he himself has repeatedly stressed that his falsifiability criterion is the criterion of demarcation between science and pseudoscience, not the meaning criterion of cognitive significance. Regardless, the term "falsification" is used to prove the negation of a proposition.

The logical form of the falsifiability criterion is as follows:

$$\{O_1, O_2, \ldots O_n\} \vdash \neg S$$

That is, a statement has cognitive significance if and only if it can be falsified by a finite sequence of observational statements. So, for instance, if we observe one instance of a swan that is not white, we can falsify the proposition "All swans are white." Thus, we can affirm that the universal statement has cognitive significance.

Here Popper skillfully uses logical asymmetry. In quantificational logic, the universal quantifier "All" and the existential quantifier "Exist" can be transformed into each other: The negation of a universal proposition can be written in the form of an existential proposition, and vice versa. This transformation can be written as follows:

$$\neg\forall x f(x) \leftrightarrow \exists x \neg f(x)$$

For example, the negation of "All swans are white" is "Not all swans are white," which is logically equivalent to the existential proposition "Some swans are not white."

Those things we can directly observe in our daily lives are usually written as particular propositions, such as Fa – i.e., ×× is a swan. Then we can have $Fa \wedge Ga$ (that is, ×× is a swan and ×× is white) or $Fa \wedge \neg Ga$ (that is, ×× is a swan but is not white).

Existential propositions can be verified by particular propositions. For instance, from $(Fa \wedge Ga)$ we can logically deduce $\exists x(Fx \wedge Gx)$ – that is, from "×× is a white swan" we can deduce that "Some swans are white." On the contrary, strict universal propositions cannot be verified by particular propositions. For example, we cannot deduce that "All swans are white" from "×× is a white swan." Because the negation of universal proposition can be written in the form of an existential proposition – for example, the negation of "All swans are white" is "Some swans are not white" – universal propositions cannot be verified, but can be falsified.

Nevertheless, the falsifiability criterion can be met with a host of counterexamples, too. The most prominent problem is that existential propositions can be verified, but their negation propositions can be written in the form of universal propositions, so many existential propositions are not falsifiable. In this way, the falsifiability criterion will exclude a large number of existential propositions as having no cognitive significance. Clearly, such a consequence is unacceptable.

For example, we usually think that the statement "There is extraterrestrial life" is meaningful, which also guides scientists to explore the universe. The statement "There is extraterrestrial life" can be verified as long as we find some form of extraterrestrial life. But this statement cannot be falsified because the universe may be infinite, and so even if we check a large number of planets and find no extraterrestrial life, we still cannot completely falsify the statement.

In addition, we can make any meaningless sentence satisfy the falsifiability criterion with the help of a conjunction. If a proposition can be falsified by a series of observational propositions, then its conjunction with any meaningless sentence N can also be falsified by those observational propositions. That is to say, $S \wedge N$ can satisfy the falsifiability criterion, so it should also have cognitive significance. But we have shown that N is a meaningless sentence, so the compound sentence containing a meaningless statement should also be meaningless.

The falsifiability criterion thus has the same problem as the verification principle. That is, universal propositions may have cognitive significance, but many of their negations, as existential propositions, cannot themselves be falsified, so they apparently have no cognitive significance.

As mentioned at the outset of this section, Popper has repeatedly claimed that his falsifiability criterion is simply the demarcation criterion between science and pseudoscience, not the criterion of cognitive significance. This is because scientific statements usually appear in the form of universal propositions, so they can be falsified. However, W. V. Quine has also raised an objection to this. He argues that the internal form of some universal propositions is so complex that they cannot be falsified. For example, science will tell us that "All men are mortal" cannot be verified if the total number of men is infinite. But how could this statement be falsified? It seems that as long as we find a counterexample, such as "×× is immortal," we can falsify the proposition that "All men are mortal." But if we want to find such a counterexample, we need to verify that "For all time, ×× is alive." This is logically equivalent to verifying a strict universal proposition, which cannot be achieved. As a result, "All men are mortal" can neither be verified nor falsified (Quine 1974, p. 219).

The confirmability criterion

A. J. Ayer was the initial advocate for the confirmability criterion, proposing that a statement is of cognitive significance if and only if, when taken together with other auxiliary hypotheses, it can render the deduction of observational propositions, which cannot be independently derived by auxiliary hypotheses alone. The logical form of the confirmability criterion is as follows:

$$\{S, H\} \vdash O_1, O_2, \ldots O_n;$$

and from H alone, $O_1, O_2, \ldots O_n$ cannot be deduced.

For example, if we want to know whether the law of universal gravitation has cognitive significance, we can combine it with a series of auxiliary hypotheses (such as the data of the Earth's mass and radius, the influence of the Earth's rotation, and that air resistance is ignorable) and calculate the distribution relationship between the distance and time when an object freely falls. The distribution data is observable, and the data cannot be derived from the above auxiliary hypotheses alone. Consequently, we infer that the law of universal gravitation is a statement of cognitive significance. On the contrary, statements such as "Reason is Substance, as well as Infinite Power" are meaningless, because we can deduce no observational proposition even when combined with auxiliary hypotheses.

Compared with the verification principle and the falsifiability criterion, the confirmability criterion allows universal statements and existential statements to be meaningful in principle. For example, universal statements such as "All swans are white" can render the deduction "×× is white" if it is taken together with the auxiliary hypothesis "×× is a swan." Similarly, from the existential statement "There is a melting point for gold," in conjunction with other auxiliary hypotheses, we can infer that when gold is heated to a certain temperature, "The pieces of gold melt." On these grounds, Ayer argued that the confirmability criterion is a more adequate criterion of cognitive significance.

Nevertheless, the confirmability criterion also has its own problems. Consider any meaningless statement N and the auxiliary hypothesis $N \rightarrow O$. From the conjunction of these two statements we can derive the observational proposition O, which cannot be deduced by the auxiliary hypothesis alone. In this way, from any meaningless sentence can be deduced at least one observational proposition by resorting to the corresponding auxiliary hypothesis, so as to meet the confirmability criterion.

To deal with the problem, Ayer adds a condition: The auxiliary hypothesis h is either an analytical proposition or can be independently verified. $N \rightarrow O$ is not an analytical proposition, nor can it be tested independently, so it cannot be an auxiliary hypothesis.

However, even the improved version of the confirmability criterion also meets with problems. Let S be a statement satisfying the confirmability criterion and N be any meaningless statement. Then the conjunction $S \wedge N$ also satisfies the confirmability criterion. In this way, meaningless sentences will still be introduced again.

Proponents of various stripes long strove to further modify and improve the verification principle in different ways, until the logician Alonzo Church constructed Church's formula, which effectively shows that any statement can satisfy the verification principle as long as it is supplemented by the formula as an auxiliary hypothesis. Church's formula is as follows (Scheffler 1964, pp. 153–154):

$$(\neg O_1 \wedge O_2) \vee (O_3 \wedge \neg S)$$

When this formula is combined with O_1, we can deduce O_3, so Church's formula can be tested separately. In other words, it can meet the requirements of the auxiliary hypothesis in the confirmability criterion. When Church's formula, as an auxiliary hypothesis, is combined with any statement S, we can deduce O_2. S can satisfy the confirmability criterion, but S can be any sentence.

Therefore, any meaningless statement can satisfy the confirmability criterion with the auxiliary hypothesis of Church's formula. As a result, Church's formula has convinced philosophers of science to give up the confirmability criterion as the criterion of cognitive significance.

The translatability criterion

Philosophers of science have tried time and again to find an adequate testability criterion of cognitive significance, but all attempts have ended up with logical difficulties of some sort. As a result, many philosophers have turned to the translatability criterion instead.

The translatability criterion of cognitive significance constructs a set of artificial empiricist language. Then, a sentence has cognitive significance if and only if it is translatable into the empiricist language. Such an artificial empiricist language L will be composed of basic vocabulary and sentence formation rules. According to Hempel:

> (a) The vocabulary of L contains: (1) The customary locutions of logic which are used in the formulation of sentences; including in particular the expressions "not," "and," "or," "if... then...," "all," "some," "the class of all things such that...," and "...is an element of class..."; (2) certain observation predicates. These will be said to constitute the basis of an empirical vocabulary of L; (3) any expression definable by means of those referred to under (1) and (2). (b) The rules of sentence formulation for L are those laid down in some contemporary logical system such as *Principia Mathematica*.
>
> (Hempel 1950, pp. 41–63)

In comparison with the testability criterion, the translatability criterion has many advantages. First, the translatability criterion can include all quantificational symbols, since both universal and existential quantifiers belong to the basic vocabulary of language L. Second, sentences without cognitive significance, such as "Caesar is a prime number," can no longer be introduced by means of logical connectives (conjunction or disjunction), for the compound sentence contains a meaningless statement that is not untranslatable. Third, if a proposition is meaningful, according to the rules of logic, its negation satisfies the translatability criterion and therefore is meaningful. Thus, the translatability criterion can avoid the problem encountered by the testability criterion, which we have shown that, through appropriate logical construction, any meaningless sentence can satisfy. The great advantage of the translatability criterion, then, is that it would not regard all sentences as cognitively meaningful, but would consider those meaningless sentences as ultimately untranslatable.

The translatability criterion seems to have many advantages. Now the main problem is how to translate, in particular, theoretical terms or dispositional terms into observable sentences. For example, glass has the property of "fragility." "Fragility" is not directly observable, but is a dispositional term that can be made visible only under specific circumstances. To translate theoretical or dispositional terms, there are the requirements of "definability" and of "reducibility," respectively.

The requirement of definability

The main requirement of definability is, simply, that all theoretical terms must be definable. For example, some scientists and philosophers try to define theoretical terms as a series of operations. P. W. Bridgman, a great American physicist, published *The Logic of Modern Physics* in 1927 and put forward the concept of "operationalism" (Bridgman made great contributions to experimental physics and was awarded the 1946 Nobel Prize for his research in high pressure physics). According to his definability requirement of operationalism, a term has cognitive meaning if and only if it can be defined by observational terms:

Theoretical term=df observational terms

Einstein, furthermore, defined the concepts of "simultaneity" and "time" through a series of operations in the theory of relativity, which can be regarded as the classic example of operationalism. In fact, it was Einstein's special theory of relativity in 1905 that influenced Bridgman's proposal of operationalism, not vice versa. In classical mechanics, Newton's concepts of "time" and "space" are not well defined. In particular, his concepts of "absolute time" and "absolute space" were heavily influenced by his religious beliefs, so Mach first criticized the two concepts from the perspective of positivism.

Einstein inherited Mach's criticism and attempted to define the concepts of "time" and "space" by a series of operations. More concretely, he proposed an operational definition of "simultaneity," resorting to which we can find a way of calculating "time" by setting in simultaneous motion a bunch of identical clocks.

The simultaneity of the same place is rather simple, since we can directly observe two events occurring at the same place and at the same time. But how to determine the simultaneity of different places? Einstein imagined that we might put two mirrors, which can reflect events, at the endpoints of two different places, A and B, with an inclination angle of 45°. Then we can observe at M, the midpoint of AB, whether two events at A and B happen at the same time. To do so, we place the same kind of clock at points A and B, so that after synchronizing their motion, the corresponding time at points A and B can be defined.

The above operations of simultaneity are completed in the same reference system. If the reference systems are different, can we still define the concept of "simultaneity"? Assuming that a stationary platform and a moving train are two reference systems, we can determine the simultaneity between two points A and B on the platform, which correspond to the two points A' and B' on the train, by the above mirror reflection of A. But if we would like to measure the simultaneity between A on the platform and B' on the train, we can choose neither the midpoint of AB, M, nor the midpoint of A'B', M', because the train is moving at a high velocity, so the reflected lights should pass at different distances. Thus, the simultaneity should be relative to the reference frame, and the same goes for the concept of time. Thereafter, Einstein proposed that time and space are relative, establishing his renowned special theory of relativity.

Operationalism has achieved great success in physics, but it has also encountered its own logical problems. For example, we can define "intelligent" with a series of operations, including that one "scores more than 120 when taking an IQ test." We then get something like this: "X is intelligent if and only if X takes an IQ test, and then X gets a score of more than 120." Its logical form is:

X is intelligent↔(X takes an IQ test→X's score is over 120)

But, we might recall, according to the definition of material implication, when the antecedent is false, the whole expression is true, regardless of whether the consequent is true or false. Therefore, if X does not take an IQ test, then the right part of the definition, "X takes an IQ test→X's score is over 120," should be true. Thereby, we must conclude that X is intelligent, according to the definition. In this way, we might conclude that all people who do not take IQ tests are intelligent, which is obviously different from our understanding of the concept of being "intelligent."

Some may thus suggest redefining "intelligent" as follows:

X is intelligent↔(X takes an IQ test∧X's score is over 120)

But the new definition will make it so that all people who do not take IQ tests are not intelligent, which is just as unacceptable as the previous conclusion. Therefore, the idea of operationalism is a great advancement, but its logical difficulties cannot be overlooked.

The requirement of reducibility

To solve the problem of the definability requirement, Carnap put forward the concept of a reduction sentence. In response to the difficulties in the operational definition of "intelligent," he proposed a different expression:

X takes an IQ test→(X is intelligent ↔X's score is over 120)

That means that if somebody takes an IQ test, then he/she is intelligent if and only if his/her score is over 120. This avoids the problem of whether he/she is intelligent when he/she does not take an IQ test. The characteristic of the reduction sentence is that it takes the operation sentence as the antecedent of the implication and takes the equivalent condition of the concept itself as the consequent.

Still, it must be pointed out that the reduction sentence gives only a partial definition. Because a complete definition of a concept must specify the characteristics displayed by the definiendum (i.e., a word, phrase, or symbol that is the subject of a definition) under all conditions, its general form should be "X is T if and only if X satisfies condition C." The reductive sentence provides definitions given under the operational conditions, so they are only partial definitions. For this reason, Carnap modifies the requirement of definability as the requirement of reducibility: Empirically meaningful terms must be obtained by means of the reduction sentences on the basis of observational terms.

With that said, the requirement of reducibility also has its own problems (as we might expect by this point). For example, "The length of $\sqrt{2+10^{-100}}$ cm" cannot satisfy the requirement of reducibility, as we cannot distinguish the length of a magnitude of 10^{-100} cm in reality. According to the requirement of reducibility, then, such a concept is meaningless. But, in fact, the length can be mathematically meaningful.

Due to the rapid development of modern science, many scientific theories have far outstretched simple observation. Therefore, many abstract theoretical concepts in mathematical and empirical sciences cannot be expressed in the form of reduction sentences. The requirement of reducibility may exclude many core theoretical concepts in modern sciences. Subsequently, many philosophers of science abandoned the requirement of reducibility as the meaning criterion and instead turned to holism.

The rise of holism

Both empirical science and mathematics need theoretical constructs. For example, in the axiomatization of Euclidean geometry, David Hilbert calls the concepts of "point," "line," and "between" as "primitive terms," with the concepts of "segment," "angle," "triangle," and "length" being "defined terms." These defined terms are constructed by the primitive terms according to the rules of the axiomatic method – i.e., by means of definitions and deductions. Similarly, abstract sciences need theoretical constructs. Many scientific concepts are to be constructed from primitive terms and basic laws. According to the model of theoretical constructs, although we cannot distinguish the quantities $\sqrt{2+10^{-100}}$ cm and 10^{-100} cm in real life, the two lengths can nonetheless be constructed within science, so they are meaningful.

If abstract sciences need theoretical construction, how can we determine whether some statement has cognitive meaning? According to the model of theoretical constructs, it is meaningless to ask whether a single statement has empirical meaning. The cognitive meaning of a given expression E depends on two factors: (1) the linguistic framework L to which the expression belongs, namely the rules of L stipulate how to derive observational sentences from a given statement; (2) the theoretical context in which E appears, for instance, the auxiliary hypotheses in L.

Hence, we can say it is meaningless to ask whether the law of universal gravitation itself has cognitive meaning. But in combination with other definitions and laws of Newtonian mechanics, it can be interpreted empirically, so it has cognitive meaning. Such an empirically interpreted system is called an "interpreted system." Cognitive meaning is not for a single sentence or term, but rather for the whole theoretical system. A theoretical system has cognitive meaning because we can interpret it by means of observational terms.

Both the requirement of definability and the requirement of reducibility are used to define or reduce theoretical terms by observational terms. According to the "interpreted system," the theoretical terms are not defined by observational terms; on the contrary, observational terms are defined by theoretical terms:

Observational term=df theoretical terms

For instance, according to modern chemistry, "gold" is defined as "the chemical element with atomic number 79." We can directly observe gold in our daily life, but "chemical element" and "atomic number" are abstract concepts that cannot be observed in the same way. Since the theoretical system of modern chemistry can explain daily experience in terms of its theoretical constructs, we say that the theoretical body of modern chemistry is meaningful.

Similarly, Carnap's reduction sentence about the concept of being "intelligent" can be transformed to interpret the empirical phenomenon "to get a score over 120" by means of being "intelligent" and "to take an IQ test":

X's score over 120$\leftrightarrow$(X is intelligent$\wedge$X takes an IQ test)

The concept of "interpreted system" solved the problems posed by the criterion of definability and the criterion of reducibility. Nevertheless, this seems to be too loose to be the criterion of cognitive meaning, for it may admit the possibility of introducing metaphysical sentences into the interpreted system. In that case, the interpreted system as a whole can be empirically interpreted, but those metaphysical sentences within it are still meaningless. K. Reach and Otto Neurath call these sentences "isolated sentences." These sentences are neither formally true nor false, nor do they have empirical content – that is, they will not affect the explanatory and predictive power of the interpreted system if omitted. How, then, can we eliminate the isolated sentences that are cognitively meaningless from the interpreted system?

With the concept of an "isolated sentence," logical empiricists attempt to modify the meaning criterion standard in the following way: A theoretical system has cognitive meaning if and only if there is no isolated sentence in the system – that is, if every sentence in the system is essential for its empirical interpretation.

However, even such correction will also be confronted with logical problems. Suppose P_1 and P_2 are observational predicates, while predicate Q is the theoretical construct in the theoretical system T, which only appears in the primitive sentences of T. With these assumptions, we can conclude that the elimination of S_1 from the system T will not affect its explanatory and predictive power, and thus S_1 constitutes an isolated sentence:

$$(S_1)\ \forall x[P_1x \rightarrow (Qx \leftrightarrow P_2x)$$

Because the formula S_1 is in the form of a reduction sentence, according to Carnap's treatment, we can regard S_1 as a partial definition of Q, which is an analytical sentence. Similarly, we may construct another isolated sentence in the form of a reduction sentence:

$$(S_2)\ \forall x[P_3x \rightarrow (Qx \leftrightarrow P_4x)]$$

Both reduction sentences S_1 and S_2 are analytic because they can be regarded as partial definitions of Q. But from the conjunction of S_1 and S_2 we can get the following expression:

$$(O)\ \forall x[\neg(P_1x \wedge P_2x \wedge P_3x \wedge \neg P_4x) \wedge \neg(P_1x \wedge \neg P_2x \wedge P_3x \wedge P_4x)]$$

Because P_1, P_2, P_3, and P_4 are all observable predicates, the sentence O is also observable and has empirical meaning. But this is obviously contrary to our existing logical knowledge: From the conjunction of the analytic propositions without empirical meaning we can finally deduce a synthetic proposition with empirical meaning!

Thus, Hempel concludes that, on the one hand, analyticity must be understood only relative to linguistic rules; on the other hand, whether or not a statement is an isolated sentence is also relative to the theoretical system, for adding S_2 may make another isolated sentence S_1 have empirical meaning.

Hempel later extended the holism criterion of cognitive meaning even further, proposing that cognitive meaning is not a dichotomy of "yes" or "no," but a matter of degree. He also proposed some features to compare different theoretical systems: (1) that the theoretical expressions must be explicit and precise, and the logical relationship between the elements and their connection with observational terms must be clear; (2) that the system must have explanatory and predictive power for empirical phenomena; (3) that the theoretical system should be formally simple; and (4) that the theories should be confirmed by empirical evidence to a certain degree. Hempel also pointed out that many theoretical systems of speculative philosophy about the universe, biology, or history are not meaningful enough, at least in comparison with modern scientific theories, so speculative philosophies are not worth further research and development.

Cognitive meaning, after all, is not a dichotomy of "yes" or "no." The principle of tolerance at work in such holism in fact forces the abnegation of logical empiricists' strict meaning criteria. As a result, some philosophers of science, such as Carnap, have regarded Hempel's analysis of cognitive meaning as the sign of the end of logical positivism (Jiang 1984, p. 32).

Empirical significance of the meaning criterion

If we finally find a successful meaning criterion to judge all statements, does the meaning criterion itself have cognitive significance? Even logical empiricists consider such a question as this. After all, the meaning criterion is not an empirical hypothesis, as it is not an empirical description of the world, but it is also not an analytic sentence or a contradictory sentence either, since we cannot deduce the meaning criterion merely from logic and syntax.

Hempel's answer to this question is that the meaning criterion is a "linguistic proposal," so it is neither true nor false. For example, we can propose that the Symposium on Philosophical Issues in Science and Technology hosted on November 9, 2002 at Tsinghua University in Beijing be called "Symposium 2002." Others can choose to accept or reject the proposal, but the proposal itself is neither true nor false. We cannot say that naming the event "Symposium 2002" is true or false, but the proposal is either good or bad.

To evaluate the linguistic proposal of the meaning criterion, we must ask whether it can fulfill two functions: whether the analysis of the definiendum is sufficient, and whether the rational reconstruction of the definiendum is successful. According to Hempel, we need to distinguish the statements with cognitive meaning from the statements without cognitive meaning in our daily lives. Although there are many problems in logical empiricists' meaning criterion, it does provide a rational analysis and reconstruction for "cognitive meaning," so the linguistic proposal is appropriate.[2]

Summary

Logical empiricists have advocated for both the rejection and proposal of various meaning criteria. Over time, the dominant meaning criterion has thus changed from the testability criterion (including the verification principle, the falsifiability criterion, and the confirmability criterion) to the translatability criterion (including the definability requirement and the reducibility requirement). However, all these outstanding efforts have encountered their own logical difficulties. Consequently, thinking with respect to the meaning criterion has taken a turn toward holism, following the suggestion that cognitive meaning is not a dichotomy of "yes" or "no," but rather a matter of degree.

The change of the meaning criterion thus became the theoretical background to the rise of holism. In fact, before Kuhn, many historians and philosophers criticized logical positivism from the perspective of the history of science and technology, yet their efforts did not play a very important role in philosophy of science. In this way, we might even say that because of the logical difficulties of logicalism, Kuhn was fortunate enough to be there at the time to do the work he did and publish his most influential book, *The Structure of Scientific Revolutions*, which gave rise to historicism. Indeed, the rise of historicism was not due to the fact that philosophy of science had previously ignored the history of science, but was rather due to the logical problems in logicalism itself, which made the emergence of historicism inevitable. Thus, to understand the history, we need to understand the logical details.

With all that said, today, the issue of cognitive meaning is, admittedly, somewhat outdated, no longer standing as a central topic in philosophy of science. In fact, logical empiricism finally abandoned the sharp dichotomy of "cognitively meaningful" and "meaningless." Nevertheless, clarity of expressions and meaningfulness of points have become the academic norm for philosophical research, which is certainly one of the great legacies of logical empiricism.

In addition, logical empiricists' courage to criticize themselves stands as a wonderful model for academics today – especially within philosophical circles in China – to follow. Philosophers of science within this school would frequently propose a certain meaning criterion, and then, in short order, find some counterexamples to overthrow those same proposals. Such a critical attitude is far more admirable than those dogmatic attitudes held by some philosophers, who seem to cling stubbornly to their own views and strive only to prove themselves right.

Notes

1 It is easy to identify analytic propositions, since they offer no description of the world. So the discussion about cognitive meaning mainly focuses on whether or not a sentence is a synthetic proposition, i.e., with empirical content. Thus, "cognitive meaning" is usually understood as "empirical meaning."

2 According to Professor Tien-ming Lee (李天命) at the Department of Philosophy, The Chinese University of Hong Kong, if everyone finally accepts the logical empiricists' linguistic proposal, the meaning criterion will be an analytic proposition, just like "All bachelors are unmarried men." In writing this chapter, I greatly benefited from Tien-ming Lee's course "Epistemology."

5 Induction and confirmation

Induction is the most frequently used method in scientific research. This chapter will first introduce some forms of inductive methods and, when misused, some related fallacies. Then we will discuss the problem of induction and the problem of confirmation in philosophy of science. Let us start with induction.

Inductive methods

Enumerative induction

Enumeration is our most common inductive method. When we wish to buy some fruit at the market, we may be allowed to try it. If the sampled fruit tastes good, we will buy it; if not, we will not buy it. That is enumeration, as the amount of fruit we tasted is only a small fraction of the total amount displayed at the fruit stand, yet we inductively generalize about the rest of fruit from the taste results. Such a reasoning process can be expressed as follows:

> The fruit I tasted is good.
> ∴All the fruits on the stand are good.

Thus the general form of enumeration is:

> The examined cases F are G.
> ∴All F are G.

Sometimes we need some statistical data, such as what might be found by calculating the proportion of males and females in a population census, or by a TV station's counting of a particular program's viewing rate. Due to the large size of a population, it is nearly impossible to calculate individually all the relevant statistics, so we would usually choose a certain number of samples and make a statistical enumeration in order to determine disparities by sex or the viewing rate of a TV program for an entire population by reasoning from the proportion of males and females or the viewing rate found in the samples. The general form of such an enumerative statistic is as follows:

In the examined cases, Z% proportion of F are G.
∴For all F, a proportion of Z% are G.

Enumerative induction has no logical necessity, and it is possible to infer an incorrect conclusion from correct premises. For example, the swans we have observed were usually white, but the enumerated conclusion "All swans are white" is still wrong, for people have in fact found black swans in Australia.

If we use enumerative induction mistakenly, we will end up with fallacies of some kind. For example, insufficient statistics may lead to the fallacy of over-generalizing. In 2002, an undergraduate student injured a bear at the Beijing Zoo, and, as a result, many Chinese newspapers lambasted the quality of all college students. This is a wildly insufficient statistic, as there are millions of college students now, and the undergraduate's personal behavior is in no way able to represent all college students. In fact, it is even unfair to accuse that single student based only on one instance of misconduct, for he may very well have done many good things, making this single behavior insufficient grounds upon which to evaluate his entire personality.

Biased statistics can lead to fallacies, too. For example, some dishonest fruit vendors often put good fruit on the top of the fruit pile and rotten fruit underneath. If we only look at the fruits on the surface, such enumeration is biased and may be very misleading.

Statistical syllogism

Sometimes we make inferences by means of statistical probability. Let us assume, for instance, that 90% of the balls in a bag are known to be white and that the rest are black balls. If we randomly pick up one ball, which color will it be? We might usually think that what we take out will likely be a white ball, which is a statistical syllogism whose form can be expressed as follows:

Z% of F are G (Z% can be replaced by "most of").
x is F.
∴x is G.

When the proportion of Z is close to 100%, the statistical syllogism forms a strong argument. But when the proportion is closer to 50%, the argument become weaker, as x has almost the same probability as non-G. When the proportion is less than 50%, we can then change the result of the above inference to "x is not G."

The argument from authority

In daily life, people often believe in authority. For instance, if we want to know what the weather will be like tomorrow, we will typically listen to the weather forecast from a weather station. Why is this the case? Here, we would likely use the argument from authority, in accordance with this general form:

> x is a reliable authority in the field of p.
> x said p.
> ∴p.

The argument from authority is very common in our daily lives, and if it were not, we would have a great deal of responsibility placed in our own hands, which would only increase our trouble and stress. For example, people rarely worry about the collapse of the building in which they live because they believe in the architect's professional abilities. When we see a doctor and take the medicine she prescribes, we also have to believe in the doctor's professional expertise and the pharmacist's professional quality. All these inferences contain arguments from authority. The argument from authority is a kind of induction, which can be analyzed in the form of a statistical syllogism:

> Most of x's judgments in the field S are true.
> p is x's judgment in the field S.
> ∴p is true.

Arguments from authority can sometimes be wrong. For example, the weather forecast is not always accurate, and architects, doctors, and pharmacists may occasionally make mistakes. However, misusing arguments from authority will also lead to fallacies. Here are some such fallacies:

(1) *Authority is misused*: For example, some philosophers use Einstein's theory of relativity to argue for moral relativism, claiming that if there is no absolute reference system in physics, then the same is true in morality. But they misuse the authority in this case, because Einstein proposed his theory of relativity with the intention of justifying the idea that all reference systems, including uniform speed and acceleration, obey universal physical laws.
(2) *The authority's judgment is outside his/her professional field*: Chairman Mao was a great revolutionary, politician, and militarist, but he was not a very good economist. During the Great Cultural Revolution, because of the prevalence among Chinese citizens of superstitious beliefs in Chairman Mao's judgments outside his professional field, China suffered a tremendous loss economically. Thus, the personality cult is in fact a kind of inductive fallacy.

(3) *Stars are used as authority*: This is a common fallacy in modern media. For example, many advertisements use movie stars. Apart from the problem of image endorsement, these stars usually do not have enough knowledge to be authorities on what they are promoting. For example, some stars advertise electrical appliances or drinks, but they are by no means electrical or nutritional experts, so their judgments are not necessarily credible. Nevertheless, many people believe in stars in the same way they trust the weather forecast – an inductive fallacy, for sure.

(4) *Some authorities do not have enough evidence to support their judgments*: For example, bishops are religious authorities, and they believe that God exists, but if their judgment is not supported by evidence, it is still not credible.

(5) *Authorities may have disagreements*: For example, in the 1990s, cold fusion theory – which holds that nuclear fusion can be realized at room temperature – was popular in physics circles. However, physicists were very conflicted as to the veracity of the theory, and many scientists disagreed with it. In the case of a non-expert, we do not know which authority to follow. Therefore, when using the argument from authority, we would be better off coming to our own conclusions only after the authorities themselves reach a consensus. Now, as it turns out, most physicists tend to think that there is no such cold fusion effect.

Arguments from anti-authority

Corresponding to the argument from authority, we can construct the argument from anti-authority, in principle. Anti-authority refers to those who are mostly wrong in a certain field. For example, Pelé was one of the greatest football players, but his predictions about football games were almost always wrong, so he can be regarded as an anti-authority. If an anti-authority makes some assertion, then we have reason to believe the negation of the assertion instead. The form of an argument from anti-authority can be expressed as follows:

x is an anti-authority in the field of p.
x said p.
∴Not-p.

The argument from anti-authority can also be expressed in the form of a statistical syllogism:

Most of x's judgments in the field S are false.
p is x's judgment in the field S.
∴p is false.

However, we can rarely find such kinds of anti-authority in reality, so the argument from anti-authority is mainly a theoretical construct and is not frequently used in reality.

On the other hand, a corresponding mistake called the "genetic fallacy" – sometimes also called "the fallacy of personal attack" – is much more common. This fallacy tries to deny an assertion made by somebody by way of criticizing his/her personality. The Lysenko Affair, which occurred in the former Soviet Union, is a famous example. In this event, the Soviet agronomist and scholar-tyrant Trofim Lysenko criticized the Austrian geneticist Gregor Mendel as a "petty bourgeois idealist," with the corresponding conclusion that Mendel's genetic theory must be wrong. This is thus to deny Mendel's theory by painting him as an anti-authority through personal attack. History, of course, has shown that Lysenko's argument was absolutely wrong.

Analogical inference[1]

Analogical inference, or the argument from analogy, is to infer some further property by comparing the shared properties of two or more things. The general form of an analogical inference is as follows:

> x kind of object has properties G, H, etc.
> y kind of object has properties G, H, etc.
> x kind of object has property F.
> ∴y kind of object has property F.

Analogical inference is frequently used in the research and development of new drugs. Since we cannot directly use humans for experiments, biologists usually do many experiments on animals first. If the new drug has a good effect on animals, the thinking is that it may have a good effect on humans as well. This is because humans and animals share some similar properties, such that the effect of drugs on animals may also apply to humans. In the context of ancient China, the theory of yin and yang can also be regarded as an analogical inference: yin and yang, just like male and female, can give birth to everything.

But analogical inference is likely to lead to mistakes. For example, there is the following wrong analogy in traditional Chinese medicine: One year has four seasons, and humans have four limbs; one year has 12 months, and humans have 12 joints; one year has 365 days, so humans should have 365 bones. Modern medicine, however, tells us that this number is wrong. As we can see, analogical inference depends on the similarity of two kinds of object, so the more similar y objects are to x objects, the more likely the analogical inference will succeed. Accordingly, in the development of new drugs, mice are usually used for experiments first, and then monkeys are used; after all, since monkeys are more similar to humans, the experimental results on monkeys should be more reliable than those from experiments on mice.

J. S. Mill's inductive methods

John Stuart Mill (1806–1873) was the son of the British philosopher James Mill. During his childhood, Mill's father educated him with utmost strictness, determined to raise him as a genius. As a result, Mill almost had a nervous breakdown, even though, in the end, he did in fact become a great philosopher. His *Principle of Political Economy* (1848) had a great influence on political economy, and his *On Liberty* (1859) and *Utilitarianism* (1861) had a lasting influence on both liberalism and utilitarianism. Furthermore, his *System of Logic* (1843), which was translated by the great thinker and translator Yan Fu (Wade–Giles romanization: Yen Fu) into Chinese and had a profound influence in China, systematically sorted out different inductive methods. In this work, Mill has five main inductive methods: the method of agreement, the method of difference, the joint method, the method of residues, and the method of concomitant variation.

The method of agreement is to find common factors in order to determine causes, and its general form is as follows:

ABCD→X
ABCE→X
ABDF→X
ACDG→X
∴A, as the only common factor, shall be the cause of X.

The result derived by the method of agreement is not necessarily true, however. To show this, Salmon gave an interesting example: Someone likes to drink alcohol. On the first day, he gets drunk when he mixes liquor with soda water; on the second day, he drinks wine with soda water and he gets drunk again; on the third day, he drinks beer with soda water and he gets drunk a third time. From this, he infers by the method of agreement that soda water is the cause of his drunkenness, when in fact the real cause of his drunkenness is alcohol.

Sometimes when some factor does not occur, the result will not appear. The method of difference is to determine the cause of the event by looking for the missing factor. Its general form is:

ABCD→X
(Not-A)BCD→Not-X
∴A is the cause of X.

The method of agreement and the method of difference can be used together, which is called the "joint method," and the general form of this is as follows:

ABCD→X (not-A)BCD→not-X
ABCE→X (not-A)BCE→not-X

ABDF→X (not-A)BDF→not-X
ACDG→X (not-A)CDG→not-X
∴A is the cause of X.

The columns are the method of agreement and the rows are the method of difference, so this is called the joint method. In modern science, we frequently use controlled experiments to find out the cause of events by systematically changing some factors, which is a use of the joint method.

Going further, there is also the method of residues. According to this method, if a series of complex factors causes some complex phenomena, and we know that some particular factors are the causes of some particular phenomena, then we can infer that the residual factor may in fact be the cause of the residual phenomena. For instance, the colors of fireworks are determined by the spectra of their constituent elements, so if we know that other colors of fireworks are produced by the combustion of some known metals, we can infer that the residual white light is generated by the residual element magnesium. The method of residues can be represented as follows:

The complex factors A, B, C, D cause the complex phenomenon a, b, c, d.
B→b
C→c
D→d
∴The residual A is the cause of a.

The method of concomitant variation means that if two successive phenomena change in proportion or inversely, we can infer that the former is the cause of the latter. For example, if a mercury column rises with an increase of air pressure, the air pressure may be the cause of the height of the mercury column. Or, when frying meat, the longer the frying time, the lighter the color of the meat will be, so the former is the cause of the latter. The general form of the method of concomitant variation is as follows:

A and X concomitantly change, either proportionally or inversely.
∴A is the cause of X.

The hypothetico-deductive method

Sometimes scientists only have very limited observational phenomena or experimental results from which to posit explanations. How, then, might one arrive at a scientific theory? Scientists can try to put forward a hypothesis first and then examine it through further observations or experiments. This is called the hypothetico-deductive method. The general form of the hypothetico-deductive method is as follows, where T represents the theoretical hypothesis and O represents the observation results:

$T \rightarrow O$, O

$\therefore T$

The hypothetico-deductive method is frequently used in scientific research. For instance, we may want to examine Hooke's law about the elastic expansion of springs. According to Hooke's law, the elongation of a spring is directly proportional to the tension. If, when a spring is subjected to a force of 5 newtons, the extension is 1 cm, then when it is subjected to a force of 10 newtons, we can derive the observational proposition that "The spring elongation is 2 cm." If we do observe that the spring extends 2 cm, then Hooke's law is confirmed, though not verified.

It is worth noting that the hypothetico-deductive method is actually a kind of induction. It is the logical fallacy of affirming the consequent if it is regarded as a deductive method, even if such inference is a very effective inductive method.

Hume and the problem of induction

With most of the inductive methods listed, we can now discuss the famous problem of induction in philosophy. A method itself is not a problem, but when we want to use that method to achieve some purpose, if it cannot be achieved well or at all, then there is a problem. The problem of induction comes from the fact that we want to determine universal and necessary laws of science through inductive methods, which by definition have no logical necessity as deductive methods. Even if all the collected evidence is true, we cannot ensure the universality and necessity of scientific laws. This is the problem of induction. Because the problem was first raised by David Hume (1711–1776), it is also called Hume's Problem.

Hume was born in Edinburgh on April 26, 1711. A great Scottish philosopher, historian, and representative of skepticism, his major works include *A Treatise of Human Nature* (1739), *An Enquiry Concerning Human Understanding* (1748), *An Enquiry Concerning the Principles of Morals* (1751), and *Dialogues Concerning Natural Religion* (1779). Moreover, his problem of induction and the "fact–value" dichotomy are still central issues in philosophical research.

Hume divided the objects of human reasoning into two kinds: relations of ideas and matters of fact. The former is the source of conceptual knowledge, which is intuitively or demonstratively certain. The latter produces factual knowledge, which cannot be justified by intuition or demonstration. For instance, we can imagine both that "The Sun rose from the East in the past" and that "The Sun will rise from the West in the future," without falling into a contradiction.

The knowledge of facts is based on causality. However, causal inference must assume that "Under similar conditions, similar causes will always produce similar effects," which cannot be proven logically. Thus, Hume argued that the inference forms of causes to effects are not determined by reason, but by custom and habit, which are general names for the principles of association (Hume 2007b, 5.1.5). And yet, such habits have no universality or necessity. For example, many people are used to brushing their teeth after getting up in the morning, but, obviously, getting up is not the universal and necessary cause of brushing one's teeth. As this shows, Hume raised a serious problem with respect to the rationality of induction.

The Chinese logician Bo Chen has reconstructed Hume's argument as follows: (1) Inductive reasoning cannot be justified by deductive logic because inductive methods extend from limited cases to infinite objects and jump from present experience to future prediction, neither of which can be guaranteed by deductive logic; (2) the validity of inductive reasoning cannot be justified by induction because using induction to justify induction will lead to infinite retrogress or circular argumentation; (3) inductive reasoning should be based on the uniformity of nature and the universality of causal laws. But these two assumptions have no empirical justification, and thus are finally due to people's habitual psychological associations (Chen 2001, p. 36).

The problem of induction is sometimes also called "Russell's turkey problem" because Bertrand Russell once restated the problem of induction in a humorous way. A clever turkey can use induction. It finds that its owner comes to feed it every morning, so it concludes that "Every morning the owner will come to feed me." But on the morning of Thanksgiving Day, the clever turkey itself is served at the table. Russell's turkey problem reminds us that when we humans use induction, we may fancy ourselves as clever as the turkey (perhaps at our own peril!).

Justifications of induction

Since inductive methods are challenged by the problem of induction, why do we continue to use induction? To answer this, we need to defend induction, and here are some of the cases to be made.

Inductive justification of induction

Some philosophers may argue that induction has been an effective scientific method so far. We have established a massive scientific system with the method, so we should continue to use induction. This is a common defense of induction. Still, the fact that this kind of justification uses induction to defend induction – that is, "Because A, A" – obviously raises the problem of circular argumentation or begging the question (Henderson 2018, 4.1).

Argument from the uniformity of nature

Some philosophers argue that there is uniformity in nature. Any regularity in nature, once discovered, will be universally true. Since induction can help us to discover these laws, induction is an appropriate scientific method.

Such an argument from the uniformity of nature seems reasonable at first. But what if we ask further how we can know whether nature has uniformity? If we must infer the uniformity of nature by means of induction, we thereby use induction to defend the uniformity of nature and then use the uniformity of nature to defend induction, which is again an implicitly circular argument.

Popper's elimination of induction

Karl Popper tried to eliminate induction from science, so as to dissolve the problem of induction. He believed that we need not use induction in scientific research, but only the method of hypothesis falsification. The method of hypothesis falsification has a similar form as the hypothetico-deductive method mentioned in the last section, but they are nonetheless two different methods of inference. The method of hypothesis falsification has the following form, where T designates the theory and O the observational statements:

$$T \rightarrow O, \quad \neg O$$
$$\therefore \neg T$$

The method of hypothesis falsification is actually the inference of denying the consequent, which is also known as *modus tollens*, so it is a deductive method, not an inductive method. Popper wished to introduce the deductive method, therefore, to exclude induction from science.

Nevertheless, Popper's falsificationist approach received much criticism from both scientists and philosophers. After all, induction is frequently used in science, and it is difficult to completely exclude it. Moreover, the method of hypothesis falsification has its own logical problems.

According to the Duhem–Quine thesis (named after Pierre Duhem and W. V. O. Quine), observational statements cannot be derived from scientific theories alone. In order to make observable predictions, we need auxiliary hypotheses in conjunction with the theories. Subsequently, the method of hypothesis falsification should have the following form:

$$(T \wedge H) \rightarrow O, \quad \neg O$$
$$\therefore \neg (T \wedge H)$$

The Duhem–Quine thesis shows that when the observation result does not conform to the prediction result, the theory is likely wrong, but it is also possible that the theory is right and merely the auxiliary hypothesis is wrong, or perhaps both the theory and the auxiliary hypothesis are wrong. Regardless, this shows that the method of hypothesis falsification cannot completely falsify a scientific theory.

Moreover, the underdetermination thesis also shows that, in principle, there may be an infinite number of theoretical hypotheses corresponding to the limited observational statements, but we do not have an appropriate method to judge which is the only truth. Because of this, the method of hypothesis falsification cannot provide the ultimate basis for theory choice in science. Induction is still essential.

Strawson's dissolving justification of induction

Next we come to Strawson's dissolving justification of induction. Peter F. Strawson (1919–2006) studied at Oxford University, taught there after 1947, and eventually became Gilbert Ryle's successor as Wayneflete Professor of Metaphysical Philosophy in 1968. A founder of the Oxford School of Ordinary Language, sometimes referred to as "Oxford philosophy," his representative works include *Introduction to Logical Theory* (1952), *The Bounds of Sense* (1966), *Logico-Linguistic Papers* (1971), and *Skepticism and Naturalism* (1985).

Following Wittgenstein's developments (in Wittgenstein's view, there are no "philosophical problems," only "philosophical puzzles," and the purpose of philosophical investigation is to solve these puzzles through language analysis; early Wittgenstein appealed to logic analysis, whereas later Wittgenstein employed language game analysis), Strawson also wished eventually to dissolve the justification of induction, arguing that to justify any method, a more basic method is required. Induction, as well as deduction, is one of the most basic methods to justify other methods, and itself cannot be justified any further.

Let us draw an analogy. For example, if we want to explain the legitimacy of some regulation of Tsinghua University, we need the regulations of Beijing City to show that the regulation of Tsinghua University conforms to Beijing's regulations. If we want to further explain the legitimacy of Beijing's regulations, we need national laws and regulations. If we want to explain the legitimacy of national laws and regulations, we need to use the Constitution of the People's Republic of China. However, the legitimacy of the Constitution cannot be demonstrated by other laws. In scientific methods, induction is as final and basic as the Constitution. So if induction is a basic method to justify other methods, it does not need justification itself.

Strawson claims that both induction and deduction are the most basic methods used to justify other methods. As a result, the justification of induction is in fact a kind of pseudo-problem and should be dissolved. While Strawson's dissolving of the justification may sound reasonable, however, the inquiry of the most basic method is still possible. For example, we may juxtapose the Chinese Constitution with the American Constitution and discuss their comparative reasonableness, even if not their legitimacy. Following lines like this, Hans Reichenbach provided his own pragmatic justification of induction, which we come to next.

Reichenbach's pragmatic justification

The reasoning form of Reichenbach's pragmatic justification is as follows: Events are either regular or irregular; for those events with regularity, induction can effectively help us to discover the laws; for those events without regularity, there is no effective method, and induction is harmless. Therefore, induction is always beneficial and harmless no matter whether the event has regularity or not.

Reichenbach also uses an analogy. For example, there may be fish or no fish in some area of water. If there are fish, casting a net may have some benefit; if there are no fish, casting a net will be useless and harmless. But since our goal is to catch fish, we should cast a net anyway. If we want to find the laws of nature, he would then argue, we should also use induction.

Reichenbach's pragmatic justification can only explain the rationality of induction, but it does not completely solve the problem of induction. Sometimes, using induction for events without regularity may cause disaster (think again of Russell's poor turkey). In fact, because the problem of induction cannot be solved thoroughly, research interest in philosophy of science mainly focuses on the problem of confirmation: Under what conditions does evidence confirm a theory? This brings us to our next discussion.

Hempel's studies in the logic of confirmation

Hempel sought to find the logic of confirmation, distinguishing confirmation and theory choice in science. The confirmation of a theory does not mean that we must accept the theory. For example, the Watergate scandal confirmed the hypothesis that "Nixon is a liar," but still many people did not accept such a conclusion. What Hempel focused on is the problem of confirmation of a theory.

Hempel felt the problem of confirmation is very important, as it helps to solve many related issues in philosophy of science, including (1) the relevant evidence in scientific methods; (2) the instance of hypothesis; (3) research on the rule of induction; (4) the standard of rational belief or warranted assertibility in epistemology; and (5) the meaning criterion of empiricism and operationalism (Hempel 1965, pp. 5–9).

Nicod's criterion of confirmation

Hempel first took the French philosopher and logician Jean Nicod's criterion of confirmation as an example. According to Nicod, for the law "All A are B," if x is A and x is B, then it confirms the law; if x is A but is not B, then it disconfirms the law.

Nicod's criterion can be written in the following simple logical form: For the universal hypothesis of conditional formula $\forall x(Px \rightarrow Qx)$, an instance confirms the hypothesis if and only if it satisfies both P and Q; an instance disconfirms the hypothesis if and only if it satisfies P but not Q; when an instance does not satisfy P, it is neutral or irrelevant.

For example, the universal conditional "All swans are white" is confirmed when a is both a swan and white, but is disconfirmed when a is a swan but not white. If a is not a swan, then a is neutral or irrelevant to the hypothesis.

Nicod's criterion can also be applied to multiple universal quantifiers, such as "All twins are similar to each other." Its logical form can be represented as: $\forall x \forall y(\text{Twins}(x,y) \rightarrow \text{Rsbl}(x,y))$. If a pair of individuals (a, b) are twins and look similar, the case confirms the hypothesis; if (a, b) are twins but do not look similar, the case disconfirms the hypothesis; if (a, b) are not twins, the case is neutral or irrelevant.

The Nicod standard seems to be in agreement with our intuition, yet it faces two major problems. First, it can only be applied to the hypothesis in the form of universal conditionals; it cannot deal with existential hypotheses, such as "There is extraterrestrial life," or with sentences containing combined existential and universal quantifiers, such as "For every person there is some day on which he/she shall die."

Secondly, it cannot deal with the logical equivalent transformations of universal conditionals. For example, the statement "All ravens are black" is logically equivalent to "All non-black things are non-ravens" or to "All ravens and non-ravens are either non-ravens or black." The logical form of the three statements can be represented as follows:

S_1: $\forall x(Rx \rightarrow Bx)$
S_2: $\forall x(\neg Bx \rightarrow \neg Rx)$
S_3: $\forall x((Rx \vee \neg Rx) \rightarrow (\neg Rx \vee Bx))$

The three sentences are logically equivalent. But for the confirmation of S_1, it should be "A is a raven and is black"; for the confirmation of S_2, it should be "A is not black and is not a raven"; for the confirmation of S_3, it should be "If A is a raven or a non-raven, then A is a non-raven or black." These three sentences are logically equivalent, but their confirmations are quite different, which constitutes the paradox of confirmation.

Nicod's criterion can be revised. One is to add the existential clause "There is P" to the universal conditional "All P are Q." Actually, the universal sentence pattern in Aristotle's traditional formulation has such an implication. After the revision, many of the logically equivalent sentences will no longer be equivalent. For example, the statements "All ravens are black and there are ravens" and "All non-black things are not ravens, and there are non-black things" are not logically equivalent.

Still, the correction will also run into several problems: (1) It will make many logical inferences invalid. For example, we usually think that "The combustion of sodium salts emits yellow light" and "A substance is not sodium salt if its combustion does not emit yellow light" are logically equivalent, but the above correction will make them unequal. (2) Many hypotheses in empirical science do not contain existential clauses during their formation processes. (3) Many universal hypotheses never contain existential clauses. For example, scientists may study the nature of the fertilized eggs of humans and apes, but this is just a pure hypothesis, which does not mean that scientists actually make them.

Another correction is to add a "field of application." For example, the field of application of "All P are Q" is P, and "All non-Q are not P" is non-Q. Since their fields of application are different, their confirmations are not the same case.

There are also some problems in such a correction, however: (1) Scientific hypotheses do not stipulate their field of application and their application can be arbitrary. For example, "The combustion of sodium salts emits yellow light" can also be used for negative results, such as "It must not be sodium salt since its combustion does not emit yellow light." (2) Scientific hypotheses may have a variety of logical transformations, the field of application of which is not determined.

Hempel himself answered the paradox of confirmation, claiming that it actually comes from the illusion of our psychological confusion. On the one hand, we confuse logical consideration and practical consideration. Our interest in hypotheses tends to focus on some objects, but hypotheses can actually discuss any objects. For example, in "All ravens are black," our practical consideration is ravens, but, logically, it also applies to all non-black objects insofar as they cannot be ravens. On the other hand, confirmation is related to our knowledge background. Consider the hypothesis "The combustion of sodium salts emits yellow light," yet if we put ice on the flame, it does not emit yellow light. Clearly, we do not regard this observation as a confirmation of the hypothesis because we already have some knowledge about the molecular structure of ice. Suppose, however, that we test some unknown mineral instead and observe that its combustion also does not emit yellow light. After careful inspection, we find it does not contain sodium. In this case, we will then have a confirmation of the hypothesis "The combustion of sodium salts emits yellow light."

The prediction criterion of confirmation

Nicod's criterion discusses whether an object constitutes a confirmation of a hypothesis. Hempel later pointed out that confirmation should be a relation between two statements: one describes the evidence and the other is the hypothesis. In this way, confirmation can be regarded as a special logical relationship between propositions, which can be defined by logical consequence and other pure syntactic concepts.

To do this, we first need to define the "language of science," which includes all hypothesis and evidence sentences. The concept of "hypothesis" can be expressed in the language of science. It can include general sentences, sentences containing quantifiers, or particular sentences referring to specific objects. The observation report can be a limited set of observational sentences or their conjunction. Observational vocabulary is those directly observable words, such as "black," "higher," "emitting yellow light when burning," etc. Observational sentences are sentences that either affirm or deny that an object has some observable properties – for example, that "A is a raven" or "B is not black." The observation report is a set of observational sentences that come from direct observation.

Some philosophers have proposed the prediction criterion of confirmation. Let H be the hypothesis and B the observation report, i.e., a set of observation sentences. The prediction criterion is then as follows: (1) B confirms H when B can be divided into two mutually exclusive subsets, B1 and B2, and B2 is not an empty set, and every sentence of B2 can be logically derived by the conjunction of B1 and H, but cannot be deduced from B1 alone; (2) B disconfirms H, when H and B are logically contradictory; (3) if B neither confirms nor disconfirms H, then B is neutral to H (Hempel 1965, pp. 26–27).

To give an example, consider the hypothesis "All metals expand after heating," the logical form of which can be written as: $\forall x[(Mx \wedge Hx) \rightarrow Ex]$. An observation report (Ma, Ha, Ea) can be divided into two parts: (Ma, Ha) and (Ea). Since the second part (Ea) can be derived from the conjunction of the first part (Ma, Ha) and the hypothesis, and cannot be directly derived from the first part, (Ma, Ha, Ea) constitutes a confirmation of the hypothesis. Similarly, the observation report (Sa, $\neg$Wa), "a is a swan but is not black," disconfirms the hypothesis "All swans are white" with the logical form of $\forall x(Sx \rightarrow Wx)$.

That being said, Hempel also pointed out that the prediction criterion of confirmation has shortcomings: The logical forms of scientific hypotheses are often complex, and many confirmations do not meet the prediction criterion. For example, for the logical formula $\forall x[\forall y R_1(x,y) \rightarrow \exists z R_2(x,z)]$, when all x and all y have the relationship R_1, we can predict that x and z have the relationship R_2. Since y may have infinite objects, the logical formula is therefore unpredictable. $R_1(a, b)$ and $R_2(a, c)$ is a confirmation of the logical formula, but this does not satisfy the prediction criterion.

The satisfaction criterion of confirmation

All the above criteria of confirmation have their own unique logical problems, leading Hempel to propose that the criterion of confirmation must meet a series of conditions. It should be applicable to various hypotheses with all kinds of logical complexity, and it should additionally meet three major logical conditions (Hempel 1965, p. 32):

(1) *The entailment condition*: Any sentence that is entailed by an observation report is also confirmed by the report.

(2) *The consequence condition*: If an observation report confirms every sentence in the set K, the report also confirms any sentence that is the logical consequence of K. The consequence condition has two special cases: (2.1) the special consequence condition: If an observation report confirms the hypothesis H, it also confirms every consequence of H; and (2.2) the equivalence condition: If an observation report confirms the hypothesis H, it also confirms every hypothesis that is logically equivalent with H.

(3) *The consistency condition*: If every observation report is logically consistent, they will be logically consistent with the set of all the hypotheses

that they confirm. From (3) we can also derive two special cases: (3.1) Unless an observation report is self-contradictory, it will not confirm any hypothesis with which it is logically incompatible; and (3.2) unless an observation report is self-contradictory, it will not confirm any contradictory hypothesis.

Hempel proposed the following satisfaction criterion of confirmation: (1) The observation report B directly confirms the hypothesis H if B entails the development of H about the set of objects mentioned in B; (2) the observation report B confirms the hypothesis H if H is entailed by the set of sentences that B directly confirms. Thus, we can define the concepts of "disconfirmation" and "neutrality": (3) The observation report B disconfirms the hypothesis H if it confirms the negation of H; (4) the observation report B is neutral to the hypothesis H if it neither confirms nor disconfirms H (Hempel 1965, p. 37).

Hempel also defined the concepts of verification and falsification according to his criterion of confirmation. Verification is a special case of confirmation – that is, B verifies H if and only if B conclusively confirms H, whose sufficient and necessary condition is that B entails H. Likewise, B falsifies H if and only if B conclusively disconfirms H, whose sufficient and necessary condition is that B and H are logically incompatible.

Carnap's inductive logic

While Hempel's approach to confirmation is mainly qualitative, many philosophers of science have attempted to give quantitative calculations of confirmation. Representative examples of this approach include Carnap's inductive logic and, popular in recent years, Bayesianism. They both study the degree of confirmation by means of probability theory, and we will treat each in turn.

The axioms of probability theory are: (1) $Pr(A) \geq 0$; (2) if A is a necessary event then $Pr(A)=1$; (3) if A and B are mutually exclusive, then $Pr(A \vee B)=Pr(A)+Pr(B)$. All systems following the three axioms also conform to probability theory. Although the axiomatic system of probability theory is fixed, there are many interpretations of probability. Here we only mention the five principal interpretations:

(1) The classical view. In the early version, probability theory assumed that the probability of the occurrence of basic events is equal, so it was also called the equiprobable view. For example, a dice has six sides, which constitute six basic events, respectively. The probability of these basic events should be equal, so they are all denoted as 1/6. Still, the classical view cannot be applicable on many occasions; for example, if a dice is filled with lead, the probability of all six sides will not be equal, just as the probability of a tossed coin landing on heads or tails will not be the same if the coin is unevenly weighted.

(2) The relative frequency view, with Hans Reichenbach and Richard von Mises as main advocates. Reichenbach's proposal was that the probability of some event is the relative frequency of that kind of event occurring when the total number of events tends to infinity. For example, the probability of the outcome 1 of rolling a dice is equal to dividing the number of occurrences m whose outcome is 1 by the total number of rolls n. When n tends to infinity, the relative frequency m/n is its probability. If the dice is even, the relative frequency of outcome 1 should be 1/6; however, if the dice is not even, the relative frequency may not be equal to 1/6. So, in Reichenbach's view, the classical view is a special case of the relative frequency view.
(3) Logical probability, represented by Carnap. In his opinion, probability theory can be used to calculate the support relationship between propositions, on the basis of which he elaborated his system of inductive logic.
(4) The subjectivist view, also known as the personalist view, developed by Frank Ramsey and Bruno de Finetti. They interpreted probability as someone's rational degree of belief about some event or proposition. Bayesianism, which will be discussed below, mainly adheres to this view.
(5) The propensity view, which has mainly been applied in quantum mechanics. Some philosophers such as Karl Popper have suggested that the probability wave of quantum mechanics can be interpreted as the disposition of microscopic particles to appear in a certain place at a certain time. For example, if the radioactive half-life of uranium 238 is λ_{238}, then it has the propensity of $1\text{-}\exp(-\lambda_{238}\times\theta)$ to decay at time θ.

Carnap's inductive logic is to interpret probability theory by the logical probability view. He pointed out that deductive logic and inductive logic are two fundamentally different kinds of logic, as deductive logic deduces logical results step by step from given premises, while inductive logic studies the degree to which premises support the conclusion, thus dealing with inference at a meta level. But according to inductive logic, we can calculate that the support degree of deductive inferences is equal to 1 (Carnap 1950).

Carnap also distinguished logical probability and statistical probability. Statistical probability is a description of natural phenomena, such as "The probability of a white ball being in my pocket is 90%," which belongs to the category of object language. Logical probability, sometimes called "inductive probability" by Carnap, studies the relationship between propositions describing natural phenomena, so it belongs to the category of metalanguage. For example, for the following inference, although both r and r' are equal to 0.9, they belong to statistical probability and logical probability, respectively:

The probability of a white ball being in my pocket is 90% (r=0.9, statistical probability).

Taking a ball from the pocket.

(r'=0.9, inductive probability)

═══════════════════════

It is a white ball.

The most important idea of inductive logic is that the degree of the support of evidence e to the hypothesis h can be calculated, and then the degree of confirmation C(h, e) can be obtained. When we have different candidate hypotheses, we can calculate $C(h_1, e)$, $C(h_2, e)$, and $C(h_3, e)$, so as to compare the hypotheses h_1, h_2, and h_3 (Schilpp 1963, pp. 966–998).

If Carnap's inductive logic were to succeed, scientists' research would become simpler: Scientists would only need to propose hypotheses and collect evidence, and then the theory choice could be completed by a computer equipped with inductive logic programming, with the result that the scientific theory with the highest degree of confirmation would be selected.

Furthermore, if inductive logic succeeded, this would also bring convenience to our daily lives, as many conclusions in life could be calculated. For example, young people often worry about whether their sweetheart loves them or not. With inductive logic, we could collect relevant evidence (such as whether the sweetheart dates you or accepts your gifts) and then propose two mutually exclusive hypotheses, H_1: "My sweetheart loves me" or H_2: "My sweetheart does not love me." Then we could calculate the degree to which the evidence supports these two hypotheses, before finally coming to a conclusion. In this way, we could have much less unnecessary worry in our lives.

Seen in this light, the idea of inductive logic is groundbreaking, and Carnap has also made outstanding contributions to the development of the system of inductive logic. However, as Nelson Goodman (1906–1998) has shown, confirmation must also involve practical consideration.

Goodman's new riddle of induction

Nelson Goodman was born in Somerville, Massachusetts on April 7, 1906. He received his B.S. from Harvard University in 1928 and his Ph.D. in 1941. After serving in the U.S. Army from 1942 to 1945, he went on to teach at the University of Pennsylvania, Brandeis University, and finally Harvard University, where he taught from 1968 until his death on November 15, 1998. His major works include *The Structure of Appearance* (1951), *Fact, Fiction, and Forecast* (1954), *Language of Art* (1965), *Problems and Projects* (1972), and *Ways of Worldmaking* (1978). Goodman received little attention in mainland China, but the truth is that many great Western philosophers, such as Rudolf Carnap and Noam Chomsky, were influenced by him.

In *Fact, Fiction, and Forecast*, Goodman puts forward the new riddle of induction. In doing so, he defines "grue," which is a combination of "green" and "blue," as follows (Goodman 1983, p. 74):

> x is grue=df x is examined before T and green or x is not so examined and blue

According to this definition, since up to this moment, before T, all the observed emeralds are green, the existing evidence for two hypotheses,

H_1: "All emeralds are green" and H_2: "All emeralds are grue," suggests that the degree of confirmation should be the same. Yet we might usually think that the evidence indicates that "All emeralds are green," not that "All emeralds are grue."

Some may argue that according to the principle of simplicity we must choose H_1, which is simpler. But such a choice actually depends on our linguistic convention. Because we can also define "x is bleen" as "x is examined before T and blue or x is not so examined and green," then it is possible for us to define "green" and "blue" by means of "grue" and "bleen" as follows:

x is green=df x is examined before T and grue or x is not so examined and bleen
x is blue=df x is examined before T and bleen or x is not so examined and grue

The new definitions will make "grue" and "bleen" simpler than "green" and "blue," so it is useless to appeal to the principle of simplicity.

Goodman concludes that it depends on which predicate we would like to "project" or predict. Because there is a pragmatic aspect to projection, confirmation is not merely logical calculation, but it also has a pragmatic dimension.

Bayesianism

Bayes' theorem

Two years after Reverend Thomas Bayes' death, the Royal Society published his paper on probability in 1763. In the paper, Bayes proposed Bayes' theorem about conditional probability:

$$P(A/B)=P(A\&B)/P(B)=P(B/A)\times P(A)/P(B)$$

The theorem can be derived from the axioms of probability theory, where A and B represent arbitrary events. We can substitute A and B with T (theory) and E (evidence), respectively, so that Bayes' theorem can be rewritten as follows:

$$P(T/E)=P(E/T)\times P(T)/P(E) \qquad \text{Formula 1}$$

In other words, the probability of the theory when the evidence is true is equal to the probability of the evidence when the theory is true, multiplied by the probability of the theory, divided by the probability of the evidence.

If we consider that our confidence in evidence or theory usually depends on our knowledge background B, then according to the axioms of probability theory, we can also determine:

$$P(T/E\&B)=\frac{P(E/T\&B)\times P(T/B)}{P(E/B)} \qquad \text{Formula 2}$$

Formula 2 means that the probability of the theory when the evidence and the knowledge background are both true is equal to the probability of the evidence when the theory and the knowledge background are both true, multiplied by the probability of the theory when the knowledge background is true, and divided by the probability of the evidence when the knowledge background is true. But what do those probabilities mean? This requires an interpretation of probability.

The subjective interpretation of probability

Frank Ramsey and Bruno de Finetti have proposed the subjective interpretation of probability. In their view, people's rational degree of belief about events or propositions should follow probability theory. Here, the so-called rational degree of belief is someone's degree of confidence about the occurrence of some event or the truth of some proposition. If his/her degree of belief does not follow probability theory, we can design a Dutch book to make him/her lose money.

"Dutch book" literally means "Netherlands gambling," but the Chinese philosopher of science Chen Xiaoping transliterates it into Chinese as "big loss gambling" (大弃赌) to show that any gamble can win or lose, but the result of a Dutch book is always loss. Ramsey proves that if anyone does not follow the axioms of probability theory, we can design some related gamble to make him/her always lose – that is, the "Dutch book argument" (Chen Xiaoping 1997, pp. 1–9).

Bayesianism and the problem of confirmation

If people's rational degree of belief follows probability theory, then we can take P(T) in Formula 1 or P(T/B) in Formula 2 as "prior probability" before doing an experiment or having evidence. When we do the corresponding experiments to obtain the relevant evidence, our rational degree of belief must follow probability theory, so we can calculate the "posterior probability" P(T/E) from Formula 1 or P(T/E&B) from Formula 2. This will help us solve the problem of confirmation.

For instance, we can judge whether the evidence E confirms the theory T by comparing the prior probability and posterior probability. Taking Formula 2 as an example, there can be three possibilities:

If $P(T/E\&B)>P(T/B)$, then E confirms T.
If $P(T/E\&B)<P(T/B)$, then E disconfirms T.
If $P(T/E\&B)=P(T/B)$, then E is neutral to T.

Similarly, Bayesianism can also be applied in theory choice. Since the theory T is either tenable or not, the sum of P(T/E&B) and P(¬T/E&B) must be equal to 1. If P(T/E&B) is greater than P(¬T/E&B), then under the conditions of the evidence E and the knowledge background B, our rational degree of belief in the theory T is higher than that of not-T, so the theory T should be accepted. If otherwise, the theory T should not be accepted.

When comparing two competing theories, T_1 and T_2, we can also make a theory choice by means of Bayesian algorithm. If the value of P(T_1/E&B) is greater than P(T_2/E&B), then the theory T_1 has a higher rational degree of belief in the case of evidence E and knowledge background B; therefore, T_1 is a more acceptable theory than T_2.

Bayesianism can calculate our subjective judgment with probability theory, so if it succeeds, then it may solve the problem of confirmation. Coincidentally, because the first names of Bayes and Kuhn are both Thomas, Wesley Salmon wrote a paper entitled "Rationality and Objectivity in Science or Tom Kuhn meets Tom Bayes" to suggest that "an appeal to Bayesian principles could provide some aid in bridging the gap between Hempel's logical-empiricist approach and Kuhn's historical approach" (Salmon 1990, pp. 175–204).

Bayesianism is currently a popular trend in Western philosophy of science and has been listed as one topic at the biennial meetings of the Philosophy of Science Association. However, there remain many controversies regarding Bayesianism. For example, there are objections to the probability laws as standards of synchronic coherence, as well as to the simple principle of conditionalization as a rule of inference, among other challenges (Talbott 2008).

Summary

This chapter has introduced the main methods of induction and then discussed Hume's question of induction. To justify induction, the author has variously mentioned Popper's elimination of induction, Strawson's dissolving of the justification of induction, and Reichenbach's pragmatic justification of induction, among others. The problem of confirmation and Hempel's solution have also been tackled, as well as Carnap's inductive logic and Goodman's criticism. Finally, the popular trend of Bayesianism was briefly introduced.

It is fair to say the problem of induction is the most central issue in general philosophy of science, for if this problem can eventually be solved, then many other issues in philosophy of science – such as demarcation, underdetermination, and relativism – can also be solved. This is also the reason why Bayesian theory has become such a hot topic in philosophy of science circles in recent years.

The author personally holds the view that the problem of induction or confirmation cannot be reduced to logical or mathematical calculations. Even if Bayesianism can solve the confirmation problem of scientific theories, it cannot solve the problem of theory choice in the sciences. Because theory choice involves a variety of value judgments, such as simplicity, fruitfulness, and so on, the issue is not simply one of confirmation. Scientists sometimes give up hypotheses with a high degree of confirmation and choose those with developmental potential or formal simplicity. Scientific research needs the guidance of methodology, true, but it is also an art of inquiry, meaning human beings' subjective initiative is always indispensable.

Note

1 Some textbooks divide reasoning methods into three kinds: deduction from general to particular; induction from particular to general; and analogy from particular to particular. This book, following Wesley Salmon, classifies the reasoning method with logical necessity as deduction, and those without logical necessity as induction.

6 Scientific explanation models and their problems

Introduction

We have to explain numerous events every day. In our daily lives, we may ask: What is really going on during a solar eclipse? Why do I have a cold today? What is the reason that China has developed so quickly in recent years? Similarly, we have many types of explanation in scientific research. For example, college students write experiment reports to explain why an experiment rendered a certain result. We might ask, then, is there any general form of explanation? If any, what is the logical form of scientific explanation?

At the beginning of human civilization, people tended to explain nature in mythological and anthropopathic ways, attributing all natural events to anthropomorphized gods. For example, when considering the question of why it rains or thunders, the ancient Chinese believed that there are dragon kings who take care of rain, and a thunder god who creates thunder. Therefore, "agents" in mythology explained natural phenomena.

Eventually, philosophers began trying to give the world a metaphysical explanation as they searched for the ultimate cause. For example, Aristotle used his four causes – Material Cause, Formal Cause, Efficient Cause, and Final Cause – to explain everything in the world. However, if we continue to ask where the ultimate form, effect, and telos come from, we notice that Aristotle might have had no other choice but to appeal to God as the ultimate cause, which means we cannot get rid of a "metaphysical agent."

Facing such a problem, some scientists, such as G. R. Kirchhoff and Ernst Mach, claimed that scientists should ask *how* instead of asking *why*. The reasoning behind this is simple: To answer a *how* question, we only need a mathematical description of nature, which avoids the "metaphysical agent" problem that arises when searching for *why*.

Starting in the 1930s, circles within philosophy of science began to focus on the general form of scientific explanations. At that time, Hans Driesch, a German philosopher and biologist, used the term "entelechy" to explain regeneration and reproduction in biology. He argued that every creature has its own entelechy, although it is invisible and even undetectable, just like an electric or magnetic field. The complexity level of entelechy increases from plants to animals. For example, the gecko can regenerate its tail after losing a previous one, and a human's injured fingers can heal themselves automatically to a certain extent. These examples are due to entelechy taking effect. Hans Driesch used the term extensively to explain most biological phenomena, and he even regarded the human mind as part of entelechy.

At the 8th World Congress of Philosophy in 1934, held in Prague, Rudolf Carnap and Hans Reichenbach both criticized Driesch, saying that his explanation introduced new terminology but did not actually offer any new scientific discoveries; thus, they argued, explanations based on entelechy were no more than pseudo-explanations. To illustrate this idea, Carnap (1995, pp. 12–19) wrote a chapter specially dedicated to discussing the general form of scientific explanation).

Afterwards, Karl Popper and Carl Gustav Hempel both offered their discussions on scientific explanation, though it is commonly believed that Hempel's discussion was conducted more clearly and completely. Because of this, let us begin with Hempel's scientific explanation models.

Hempel's scientific explanation models

The DN model of scientific explanation

In 1948, Hempel put forward the Deductive-Nomological model of scientific explanation, which is also called the DN model. The structure of the DN model can be shown as follows:

$C_1, C_2, \ldots, C_k$	Initial conditions
$L_1, L_2, \ldots, L_r$	General laws
E	Description of empirical phenomenon to be explained

In this model, C is the initial condition, while L is a general law, with both constituting the explanans. We can deduce E from the conjunction of L and C, which means the explanandum E is a logical consequence of the explanans.

Hempel provides an example about a frozen crack on a car radiator. The initial conditions are as follows:

(1) The car has been outside for the whole night.
(2) The outside temperature was below 25°F, and atmospheric pressure was normal.
(3) The most pressure that the car radiator could bear was P_0.
(4) The radiator was full of water and sealed up.

General laws include:

(1) The freezing point of water is 32°F under normal atmospheric pressure.
(2) When the temperature is below freezing point and the volume is unchanged, the pressure of water will rise as the temperature goes

down. There is thus a function relationship between temperature and pressure.

From these initial conditions and general laws, we can calculate the pressure P of water stress on the radiator, which is larger than the maximum pressure that the radiator can bear, P_0. Therefore, we can logically deduce that the radiator cracked, which is the empirical phenomenon to be explained.

Hempel claims that the DN model must satisfy three logical conditions and one empirical condition. Logical conditions are as follows:

(1) Explanandum must be the logical consequence of explanans. In other words, the explanandum must be logically derived from the information of the explanans; otherwise, the explanans are not sufficient to explain the explanandum. The aim of the condition is to ensure that the correlation between explanans and explanandum is inevitable, not accidental. If we can logically reach the explanandum from the explanans, the explanandum must be true when the explanans are true. This condition is also called the "deductive thesis."
(2) Explanans must include general laws, and those laws are necessary when deriving the explanandum. The inclusion of general laws is to ensure that the derivation of the explanandum from the explanans is replicable and also regular. This condition is also called the "covering law thesis." Certainly, explanans usually also includes descriptions, which are initial conditions and not lawlike.
(3) Explanans must include empirical contents, which means that at least the explanation can be tested by experiments or observations in principle. Therefore, Driesch's entelechy is excluded from scientific explanations of life phenomena, since entelechy cannot be tested through experimentation or observation.

The DN model needs to satisfy an empirical condition as well: Statements consisting of explanans must all be true. If general laws or initial conditions are false, it is not a scientific explanation, even if we can derive the explanandum from the explanans.

The IS model of scientific explanation

Dealing with statistical explanations in scientific research, in 1962 Hempel devised the Inductive-Statistical model (or IS model), based on the DN model. The structure of the IS model is as follows:

Fi	Initial conditions
p(O, F)=r (r is close to 1)	Statistical law
O_i	Phenomenon to be explained

For example, take the condition "I feel a cold breeze after sweating" as an initial condition. People who feel a chill after sweating do not always get a cold, but they do have a higher chance (maybe 80% or higher) of being attacked by a cold. Therefore, the law we have is a statistical one: Feeling a chill after sweating means there is an 80% possibility of catching a cold. The conjunction of initial condition and statistical law supports the explanandum "I caught a cold" to a high degree, so it is a scientific explanation.

It is noteworthy here that we can logically derive "I have an 80% possibility of catching a cold" from initial conditions and statistical laws. For this sort of inference, Hempel coined the term "Deductive-Statistical model" (DS model) (Hempel 1965, pp. 380–381). The logical form of the DS model is as follows:

F_i	Initial conditions
$p(O, F)=r$ (r is close to 1)	Statistical law
$p(O_i)=r$	Phenomenon to be explained

The DS model only shows us the possibility of a specific event, however, such as "I have an 80% possibility of catching a cold," instead of a certain event, such as "I caught a cold." Because of this, Hempel focused more attention on the DN model and the IS model.

Based on the logical form of the IS model, we can only logically derive "I have an 80% possibility of catching a cold" without reaching the explanandum "I caught a cold." Therefore, with the IS model, the explanans gives the explanandum a high level of support, but the explanandum is not inevitable. The inference here is inductive, not deductive. Accordingly, there are two lines between explanans and explanandum in the IS model in order to show the differences when contrasted with the deductive derivation of the DN model (one line between explanans and explanandum).

The IS model must satisfy three logical conditions and two empirical conditions. The logical conditions are as follows:

(1) The explanandum must have a high possibility of being derived from the explanans.
(2) The explanans must include at least one statistical law that is necessary for derivation of the explanandum.
(3) The explanans must include empirical content, which means, at least in principle, that the explanation can be tested through experimentation or observation.

To continue, the empirical conditions are thus:

(4) Statements of explanans must be true.
(5) Statistical laws of explanans must satisfy the requirement of maximal specificity (RMS).

The conditions (1)–(4) are basically the same as the conditions of the DN model. The fifth condition of the IS model is to choose the maximal specific sample. For example, "Jack passed out after eating a pound of candy." If we were to use the statistical law that people may pass out after eating a pound of candy with a possibility lower than 1 in 10,000, this would be too minimal for a satisfactory explanation. However, if Jack is diagnosed with diabetes and there is a law that people suffering from diabetes have a 99% probability of passing out after eating a pound of candy, then we have the maximal specificity of Jack's case and have therefore explained his passing out successfully.

Hempel proposed both the DN model and the IS model of scientific explanation, yet there are some notable differences between the two models:

(1) Laws of the DN model are general deterministic laws, while the IS model usually uses statistical laws.
(2) The inference of the DN model is deductive, so its results are inevitable, logical derivations. The inference of the IS model is inductive, which means the reliability of inferences is determined by the probabilities of laws. Even if the explanans are true, the explanandum may not happen as well.

On the other hand, Hempel points out that the DN and the IS models share a similar form: Both include scientific laws, which are essential to explanations. Scientific explanations must include scientific laws, so the requirement is called the "covering law thesis." Hempel (1965, p. 412) also dubbed his models "covering law models."

Supplementary specification of scientific explanation

The covering law thesis is the essential requirement of Hempel's scientific explanation models. But what are scientific laws? To clarify, Hempel distinguished scientific laws from accidental generalizations. For example, the statement "All metals are electrically conductive" is a scientific law. In contrast, "All coins in my pocket are made of nickel" is a true, universal statement, but we would never regard it as a scientific law.

If we cannot distinguish scientific laws from accidental generalizations, some explanations will be ad hoc. For example, to explain why the pen in my pocket is electrically conductive, we can use the law "All metals are electrically conductive," or we can say that "All things in my pocket are electrically conductive." Obviously, of the two, the latter is not a scientific explanation.

Hempel attempted to summarize the general characterization of lawlike sentences so that scientific laws are those descriptions that are both lawlike and true, but soon he found that we cannot summarize the logical form of lawlike sentences. For example, no one has ever seen a gold block heavier than 1 million tons, so we can make the generalization that "All gold blocks are lighter than 1 million tons." But is this proposition a scientific law? The answer is based on our understanding of the world, for if in the future scientists find that gold decomposes once its weight reaches 1 million tons, then the generalization can be considered a scientific law; otherwise, it is only an accidental generalization. Therefore, what is considered scientific law relies heavily on scientific research and is not only determined by logical analysis.

In addition, there is a "problem of ambiguity" when it comes to statistical explanation. For example, most of us (99% or more) will live more than five years after turning 30 years old. According to this statistical law, we can conclude that Jack will live at least five years after his 30th birthday. However, patients with terminal lung cancer have a high chance (96% or more) of dying within five years. Based on this new statistical law, we can make the inference that Jack, who is a patient with terminal lung cancer, will not live for more than five years. So, after all these inferences, how long can Jack live? Different explanations will give us different results, and this shows the problem of ambiguity of statistical explanation. The logical form of the problem is as follows:

$$\begin{array}{ll} \text{Argument 1} & \\ P(G/F)=r & \\ Fa & \\ \hline\hline & [r] \\ Ga & \end{array} \qquad \begin{array}{ll} \text{Argument 2} & \\ P(\neg G/H)=r' & \\ Ha & \\ \hline\hline & [r'] \\ \neg Ga & \end{array}$$

Hempel argued that statistical explanations rely on our background knowledge (such as whether or not Jack has terminal lung cancer) and are therefore not as objective as the DN model. He calls this "epistemic relativity." In order to avoid the problem of ambiguity, we need to introduce a "requirement of total evidence" into statistical explanation, which means all evidence about the explanandum should be demonstrated.

Variations of scientific explanations

Hempel was quick to remind others that, even in the natural sciences, not all explanations can completely fit into his DN, IS, or DS models. He further suggested that there are numerous variations of explanation models in scientific practice, whether for convenience or other reasons, and then he proposed three particular forms: elliptic explanation, partial explanation, and explanation sketch (Hempel 1968, pp. 62–64).

Elliptic explanation entails omitting laws and initial conditions that everyone knows in order to form a simplified explanation. If we were to add those omitted initial conditions and laws, the complete explanation would still fit into the DN or IS model. For example, when we try to explain why copper is electrically conductive, sometimes we could simply say, "Because copper is metal." The explanation omits the law that "All metals are electrically conductive," which is a well-known law for nearly everyone. If the law is added into the explanation, the complete explanation will be: "All metals are electrically conductive, and copper is a metal, so copper is electrically conductive," which apparently satisfies the DN model. Indeed, sometimes we use "All metals are electrically conductive" to explain why copper is a conductor, omitting the fact that "copper is a metal" from the explanation.

In a partial explanation, the explanandum is only a subset of the conclusion, which is derived from the explanans. For example, according to a certain psychological law, people tend to lose things when they are extremely depressed. However, we are not able to explain precisely or predict what items will actually go missing. From the initial condition "Jack is extremely depressed" and the relevant psychological law, we can only explain "Jack lost something" instead of "Jack lost his wallet." However, "Jack lost his wallet" is a subset of "Jack lost something," which thus constitutes a partial explanation.

In some explanations, laws and initial conditions are too complicated to be stated exactly, so we can only give an outline of an explanation or a sketch for the explanandum. An explanation sketch is different from a pseudo-explanation because it requires researchers to do more empirical work to fill out its content. This is because an explanation sketch offers an empirical hypothesis, which can be verified or falsified in principle.

For example, the global financial crisis in 2008 involved many factors, and the relevant laws of economics were very complicated and difficult to describe. However, we can try to explain the crisis in this way: "The United States had a subprime mortgage crisis in 2007; the economy of the United States has great influence on the global economy; this finally resulted in a global financial crisis in 2008." This offers a brief explanation outline or sketch for the global financial crisis, but as it demonstrates, although explanation sketches are apparently useful, they often lack details.

Problems of scientific explanation

Explanation and prediction

Hempel (1965, p. 367) argues for the structural identity of explanation and prediction:

(1) Every adequate explanation is potentially a prediction.
(2) Every adequate prediction is potentially a explanation.

Since scientific explanation is akin to argumentation, this means the explanandum can be deduced or induced from the explanans. If the explanandum is known, the argument is an explanation; if the explanandum is unknown, the argument is a prediction.

Nevertheless, many philosophers have criticized the structural identity of explanation and prediction. Michael Scriven, for example, has given the argument of a syphilitic mayor: Imagine that Jones, the mayor of a small city, suffers from some sort of paresis. Since the paresis usually is caused by syphilis, and Jones has been a syphilitic for many years, his longtime syphilis thus explains his paresis. However, the probability of paresis among untreated syphilitics is low (roughly 10%, meaning about 90% of syphilitics are not susceptible to paresis). With this in mind, if we use the statistical syllogism, the prediction will be that "Jones won't get paresis," so this argument of the syphilitic mayor shows that explanation and prediction have different results and that there is no structural identity (Scriven 1959).

Scriven and other philosophers have also pointed out that the theory of evolution can explain, but cannot predict, species variation. For example, we can explain why leopards run so fast and why giraffes have such long necks with the theory of evolution; these are results of natural selection. However, the theory of evolution cannot tell us what future species will look like. Therefore, the theory of evolution can explain but cannot predict (Scriven 1962).

In addition, Scriven claims that some events can only be explained afterwards but cannot be predicted beforehand. For example, we usually investigate the causes of an accident to explain a bridge collapse after it happens. We cannot predict the bridge collapse in advance. Aviation accidents often result in heavy casualties, so if we could predict accidents in advance, tragedies would be avoided efficiently. However, what we can do in reality is simply give conclusive explanations for disasters. In such events, explanations and predictions are not identical (Scriven 1963).

With that said, Hempel did provide some counterarguments to these criticisms. Concerning the argument of the syphilitic mayor, Hempel held that since the probability of paresis among untreated syphilitics is low, "syphilis causing paresis" is not a good IS explanation (a good IS explanation has to satisfy the high-probability requirement). He also believed that Darwinism offers a partial and statistical explanation of the evolution of species. Since mutations and environmental factors to a great extent are random and complex, biology cannot predict specific new species, but predictions and explanations are still identical. Concerning explanations after accidents, Hempel argued that if we could have known every specific initial condition in advance, accidents would be not only explicable but also predictable.

Asymmetry thesis

The structural identity of explanations and predictions is sometimes also called the "symmetry thesis." However, numerous philosophers of science have pointed out that many scientific explanations are asymmetrical – that is, although events A and B are regularly connected, A can explain B, while B cannot explain A. For example, according to the principles of geometry and of optics, the length of a flagpole has a proportional relation with the length of its shadow at a given time of day. Therefore, we can calculate the length of its shadow by measuring the length of the flagpole, which also can be inferred from the length of the shadow. However, we usually believe that the length of the flagpole can explain the length of its shadow, but not vice versa. Therefore, an explanation is not always symmetrical.

Similarly, according to classical mechanics, there is a correlation between the length of a single pendulum and its period:

$$T = 2\pi\sqrt{l/g}$$

We can calculate the period according to its length, but we can also calculate the length according to its period:

$$l = gT^2/4\pi^2$$

However, we usually say that the length of a pendulum explains its period; few of us believe that the period of a pendulum explains its length.

Asymmetry is not just temporal: Some events that happen in advance cannot fully explain the events that happen afterwards either. For example, drastic changes in barometric readings usually indicate storms in the future, but we would never think that changes in barometric readings can successfully and sufficiently explain the storms coming afterwards.

The asymmetry thesis is closely connected with the structural identity of prediction and explanation. Therefore, all the examples described above can also be used to demonstrate differences between explanation and prediction.

The irrelevance objection

David-Hillel Ruben, a philosopher of social science at the University of London, has proposed the irrelevance objection to Hempel's models of scientific explanation. Here is one example he borrows from Ardon Lyon (Ruben 1990, p. 182):

(a) All metals are electrical conductors.
(b) All electrical conductors are subject to gravitational attraction.
(c) All metals are subject to gravitational attraction.

Although Lyon's example satisfies all the requirements of the DN model, it is not an explanation of "All metals are subject to gravitational attraction" because gravitational mass is the true reason, and electrical conductivity has nothing to do with gravitational attraction. Thus, electrical conductivity is irrelevant information, which Hempel's explanation models cannot effectively exclude.

Peter Achinstein (1983, pp. 168–171) gives another example:

(1) Jones ate a pound of arsenic at time t.
(2) Anyone who eats a pound of arsenic will die within 24 hours.
(3) Jones died within 24 hours after t.

This example satisfies the DN model as well; however, Jones did not actually die of arsenic poisoning. He was so unfortunate that he died of a car accident soon after eating arsenic, which means that eating arsenic was irrelevant to his death, even if it could have killed him later. Hempel's explanation models cannot successfully deal with such an occasion.

Timothy McCarthy (1977, pp. 159–166) has also proposed a formula to criticize Hempel:

$$\forall x(Ax \rightarrow Bx)$$

$$C(e) \wedge A(o)$$

$$\neg B(o) \vee \neg C(e) \vee D(e)$$

$$D(e)$$

Since the first of the explanans can be a universal law, and the conjunction of explanans can lead us logically to derive the explanandum, McCarthy's formula satisfies the DN model in every aspect. However, it may not be a good scientific explanation. Considering the following example, such an "explanation" is, in fact, seen to be quite ridiculous:

All metals conduct electricity.
The forest was struck by lightning, and this watch is made of metal.
This watch does not conduct electricity, or the forest was not struck by lightning, or the forest caught fire.

The forest caught fire.

All of these examples constitute irrelevance objections to Hempel's models of scientific explanation.

Requirement of maximal specificity

Hempel also suggested the RMS for the IS model, but the RMS was specifically questioned by Wesley Salmon. To grasp his challenge, consider that salt has a high probability (say 95%) of dissolving in cold water within five minutes. We can say a "dissolving spell" has been cast over the salt, making it into "hexed salt," and then we can pose a lawlike generalization: Hexed salt has a high probability of dissolving in cold water within five minutes.

Now we need to explain a new event, namely that some hexed salt dissolved in water. According to the RMS, we should make it specific that the salt is hexed salt, and thus the explanation is as follows:

Hexed salt has a high probability of dissolving in cold water within five minutes.
Hexed salt was put into water.

Hexed salt dissolved within five minutes.

Of course, this is not a satisfactory scientific explanation. Therefore, Salmon claimed that Hempel's RMS should be corrected by the requirement of the maximal class of maximal specificity. Since both hexed and normal salt have the characteristic of dissolving within five minutes, we should choose the maximal class of maximal specificity, "salt," instead of the smaller class, "hexed salt."

However, Salmon also argues that this correction is of no help, for we can hardly choose the proper "maximal class of maximal specificity." For example, salt and baking soda both have a high probability of dissolving in cold water within five minutes. Consequently, when we explain the dissolution of hexed salt, should we choose "salt and baking soda" as the maximal class of maximal specificity? To do so would be to explain the dissolution of hexed salt with the statement "Salt and baking soda both have a high chance of dissolving in cold water within five minutes." This is certainly inconsistent with our scientific intuition. Thus, because Hempel's explanation models have been met with a host of questions and challenges, many philosophers of science have since attempted to create new conceptions and approaches for scientific explanation.

Van Fraassen: Pragmatics of scientific explanation

Hempel emphasized the logic of explanation, which usually pays no regard to context. In other words, if the DN and IS models hold, then they can hold in any context. Yet, at the same time, Bas van Fraassen's pragmatics of explanation is focused on the context of explanation.

In van Fraassen's (1977, pp. 143–150) view, traditional explanation models present three ideas: (1) Explanation is a relation simply between a theory or hypothesis and the phenomena or facts; (2) explanatory power cannot be logically separated from certain other virtues of a theory, notably truth or acceptability; and (3) explanation is the overriding virtue, the end of scientific inquiry.

Van Fraassen argues against the idea that scientific explanation, truth, and acceptance of a theory are equivalent. In his opinion, scientists accept a scientific theory simply because the theory is empirically adequate – that is, to save the phenomena. Therefore, the acceptance of a scientific theory does not need to rely on an acceptance of its truth, let alone an explanation of all phenomena in the area. Scientific explanation is neither the overriding virtue nor the end of scientific inquiry. For instance, if we have to explain the probability phenomena in quantum mechanics, we might need to introduce hidden variables; however, doing so will also bring with it certain metaphysical baggage.

Van Fraassen thinks that traditional explanation models have two prejudices. First, philosophy of science must indicate the sufficient conditions and necessary conditions concerning why theory T explains phenomenon E. Second, explanatory power is a virtue of a theory itself (or its relation with the world), such as simplicity, truth, empirical adequacy, etc. His own suggestion, then, was that:

> scientific explanation is not a (pure) science but an application of science. It is a use of science to satisfy certain of our desires; and these desires are quite specific in a specific context, but they are always desires for descriptive information. (van Fraassen 1980, p. 156)

Therefore, a successful scientific explanation is usually an appropriately successful description with information, and the truth and acceptance of a scientific theory are irrelevant.

In van Fraassen's (1980, p. 134) view, an explanation is an answer to a why-question; therefore, a theory of explanation has to be a theory of a why-question. Sylvain Bromberger (1966, pp. 86–108) originally studied the why-question, and van Fraassen later developed this line of research further. A why-question always begins with a why – for instance, "Why did Adam eat the apple?" However, in a different context, the meaning of a why-question may be different. For instance, "Why did Adam eat an apple?" can have the following three meanings in different contexts:

(1) Why did Adam (not other people) eat an apple?
(2) Why did Adam eat an apple (not another fruit)?
(3) Why did Adam eat (not play with) an apple?

Therefore, a why-question should include not only a topic (the meaning, P_k, represented by the question itself), but also a contrast-class X. A contrast-class indicates why P happened, not other cases in the contrast-class X. For instance, when considering the question "Why did Adam eat the apple?" if what we care about is why Adam ate an apple instead of another fruit, its contrast-class in the context shall be:

(1) Adam ate a banana.
(2) Adam ate a pearl.
(3) Adam ate an orange…

Furthermore, this topic and its contrast-class also have a "relevance relation." For instance, "Adam's favorite fruit is apples" and "Adam ate an apple (not a banana or an orange)" have a relevance relation, but they have no relevance relation with the statement "The solar system has nine or eight planets."

To summarize, the why-question Q expressed by an interrogative in a given context is determined by three factors:

(1) The topic, P_k.
(2) The contrast-class, $X=\{P_1, \ldots, P_k, \ldots\}$.
(3) The relevance relation, R.

Preliminarily, we may identify the abstract why-question with a triplet consisting of these three:

$$Q = \langle P_k, X, R \rangle$$

With this, a proposition A is considered relevant to Q exactly if A bears relation R to the couplet $\langle P_k, X \rangle$.

We must now discern what the direct answers to this question are. To begin, let us inspect the form of words that will express such an answer: P_k in contrast to (the rest of) X because A. Therefore, the why-question presupposes exactly that (1) its topic is true; (2) in its contrast-class, only its topic is true; and (3) at least one of the propositions that bears its relevance relation to its topic and contrast-class is also true. B is a direct answer to question $Q=\langle P_k, X, R \rangle$ exactly if there is some proposition A such that A bears relation R to $\langle P_k, X \rangle$, and B is the proposition that is true exactly if (P_k; and for all $i \neq k$, not P_i; and A) is true (van Fraassen 1980, pp. 143–145).

For instance, "Adam's favorite fruit is apples" makes clear that P_k ("Adam ate an apple") happened, while other cases in contrast-class X (such as "Adam ate a banana" and "Adam ate an orange") did not happen. So, it bears the *relevance relation* R with the topic P_k and *contrast-class* X. "Adam's favorite fruit is apples" answers the question "Why did Adam eat an apple?" in this context and is therefore a successful explanation.

Thus, van Fraassen's critique has charged that the discussion of scientific explanation is wrong from the beginning. Traditional opinion holds that scientific explanation describes the relation between theory and fact, but it is actually the relation between theory, fact, and context. A scientific explanation is an answer to a question and the demand for needed, relevant information. The question "Why did P happen?" has different meanings in different contexts. Therefore, the answer or explanation shall be different accordingly.

Having clarified the pragmatics of scientific explanation, van Fraassen (1977, pp. 143–145), taking a rather pragmatic viewpoint, declared: "Explanation is indeed a virtue; but still, less a virtue than an anthropocentric pleasure."

Salmon: Causality and explanation

Wesley C. Salmon was a well-known American logician and philosopher of science. At UCLA, under Hans Reichenbach, Salmon received his Ph.D. in philosophy in 1950. He then taught at UCLA, Washington State

University, Northwestern University, Brown University, Indiana University, and the University of Arizona. Salmon left the University of Arizona to join the Department of Philosophy at the University of Pittsburgh in 1981, and he finally retired in 1999. His main publications include *Logic* (1963), *The Foundations of Scientific Inference* (1967), *Statistical Explanation and Statistical Relevance* (1975), *Scientific Explanation and the Causal Structure of the World* (1984), *Four Decades of Scientific Explanation* (1990), and *Causality and Explanation* (1998).

In his writings, Salmon criticizes Hempel's explanation models from the relevance perspective. For instance, we can construct a counterexample of the DN explanation model as follows:

Every man who regularly takes birth control pills avoids pregnancy.
John took his wife's birth control pills regularly during the past year.

John avoided becoming pregnant.

This explanation fits the DN explanation model well, but we would never believe that it is a good explanation. The information concerning whether John took birth control pills is irrelevant because a man can never be pregnant. This example shows that relevance is an essential factor in scientific explanation.

Salmon also criticizes the high-probability requirement of Hempel's IS explanation model and proposes that statistical relevance, rather than high probability, is the key to statistical explanation. He gives the following example:

Most people who have a neurotic symptom of type N and who undergo psychotherapy experience relief from that symptom.
Jones had N-type neurosis and accepted psychotherapy treatment.

═══════════════════════════════ (r)

Jones' N-type neurosis has been relieved.

According to Hempel's IS model, if there is a high probability of r, then this is a good statistical explanation. But since a certain proportion (r') of neurotic patients will automatically recover without any treatment, Jones' recovery is not necessarily due to the treatment but potentially to natural causes. Therefore, whether the psychotherapy treatment is the relevant key of the explanation is not determined by the strength of the probability of r, but by whether r is greater than r' – which is to say, whether psychotherapy treatments increase the proportion of recoveries.

Salmon proposes statistical relevance as a kind of probability calculation. Its definition is the following: Given the condition A, the factor C is statistically relevant to the factor B if and only if $P(B/A\&C) \neq P(B/A)$. For instance, if the probability of Jones' recovery can be changed by psychotherapy treatment, then it is statistically relevant. If not, then the psychotherapy treatment shall be statistically irrelevant.

Finally, Salmon (1998, pp. 302–319) proposes five conclusions: (1) We must put the "cause" back into "because." Although some explanations are not causal, scientific explanation must demonstrate a cause. Due to the asymmetry of causation and time, scientific explanation is asymmetric too: The cause explains the effect; an earlier event can explain a later event, but not vice versa. (2) A high probability is neither the sufficient condition nor the necessary condition for scientific explanation. The key is whether the explanans are statistically relevant – that is, whether the explanans can increase the probability of the appearance of the explanandum. (3) Hempel's "principle of essential epistemic relativity" of the IS explanation model must be given up. Salmon introduces the concept of an "objectively homogeneous reference class" and suggests that a statistical explanation has the same objective of correctness as the DN model. (4) The theory of scientific explanation should include the pragmatics of explanation. (5) We should not pursue the formal model of scientific explanation, which is universally applicable to all sciences; instead, we should observe the real form of explanation in specific sciences.

However, Salmon also acknowledges that there are three controversies in his account of scientific explanation: (1) Some trouble arises when discussing the nature of laws of nature because there is no universally acceptable criterion to distinguish between scientific laws and accidental generalizations. (2) Can there be any statistical explanation of specific events? Some philosophers think explanation can only be deductive, so we can only logically deduce statistical description from statistical laws, but we cannot statistically explain a certain event. Put simply, they only accept the DS model, not the IS model, on the grounds that we can only explain "how the world works," not "what happens." (3) Philip Kitcher distinguishes between two approaches to scientific explanation: "bottom-up" and "top-down," also called "local" and "global." Both Hempel's and Salmon's explanation models are bottom-up, or local, while Kitcher prefers the top-down approach. Either way, Salmon believes that this is still an unsolved issue of scientific explanation.

Explanation: Global and local

It was Michael Friedman who first drew a distinction between the global and local forms of explanation. Friedman's proposition is that scientific explanation is a global, rather than local, understanding provided by science, which is represented in the simplification and unification of our world image. He writes:

> On the view of explanation that I am proposing, the kind of understanding provided by science is global rather than local. Scientific explanations do not confer intelligibility on individual phenomena by showing them to be somehow natural, necessary, familiar, or inevitable. However, our over-all understanding of the world is increased.
>
> (Friedman 1974, pp. 5–19)

Philip Kitcher further distinguishes between bottom-up and top-down approaches to scientific explanation. The bottom-up approach, he determines, is a kind of local explanation, which tries to derive superficial phenomena from fundamental scientific laws, while the top-down approach is a kind of global explanation, which tries to provide a global understanding of phenomena.

Kitcher's criticism is that traditional scientific explanation models are all local, and he advocates that scientific explanation should be global: To explain something is to put it into the global pattern. For this reason, he raises the concept of an "explanatory store." Given every phenomenon K to be explained, if E(K) is the most likely demonstration set to unify K, then E(K) constitutes the explanatory store for K.

Kitcher takes Newtonian mechanics as an example in the history of science, as there are many phenomena – such as planetary motion, tides, free-falling objects, etc. – that demand explanation. Since the law of universal gravitation and Newton's three laws of motion constitute a schematic sentence, the mathematical methods of Newtonian mechanics also provide an argument pattern. Thus, these complex phenomena can be unified by Newtonian mechanics. In this sense, Newtonian mechanics is a system that constitutes the explanatory store of movement phenomena.

Kitcher thinks that his "global explanation" can successfully avoid the asymmetry problem, the irrelevance objection, and the "genuine law–accidental generalization" distinction in traditional explanation models. The asymmetry problem is due to the fact that some scientific laws have logical equivalence; hence, the effect can be derived from the cause, and vice versa. But the global explanation searches for the bottom-up, global understanding of empirical phenomena, which has direction; therefore, it can avoid the asymmetry problem.

His global explanation can also exclude the irrelevance objection. Since the global explanation provides a kind of argument pattern, even if we use the irrelevant information to explain the phenomena only once, the argument pattern probably cannot be used later; therefore, the final explanatory information must be relevant.

Because Hempel failed to distinguish scientific laws from accidental generalizations, Kitcher's answer to the problem is that laws or causal relations are determined by their place in the simplest but broadest system of scientific theory. Those that can be found in the global pattern are scientific laws. Otherwise, it is merely a case of accidental generalization. Kitcher (1981, pp. 507–531) finally uses one slogan to conclude his scientific explanation model: "Only connect."

The DNP model of scientific explanation

Peter Railton received his Ph.D. in 1980 from Princeton University, where he chose scientific explanation as his research project, and his Ph.D. dissertation was titled "Explaining Explanation." His later major research has focused on ethics, especially metaethics. Now he is a professor at the University of Michigan and a fellow of the American Academy of Arts and Sciences.

Specifically, Railton has proposed the Deductive-Nomological model of Probabilistic Explanation (or the DNP model). The core idea of Railton's DNP model is that scientific explanation must clarify the internal mechanism: To seek explanation is to seek an internal mechanism. In his opinion, if the world is a huge machine, then the purpose of scientific explanation is to search for its structure and working mechanism, which shall be more important than prediction or controlling effects. He proposes the DNP model in the following form (Railton 1978, pp. 206–226):

(a) To derive the statistical law (b) from scientific theories	Theoretical derivation
(b) $\forall t \forall x[Fx,t \rightarrow P(Gx,t)]=r$	Statistical law
(c) Fe,t_0	Initial conditions
————————————	Deductive reasoning
(d) $P(Ge,t_0)=r$	Event propensity
(e) Ge,t_0	Parenthetic addendum

For instance, we may want to explain why, at t_0, e has property G, which can be represented as Ge,t_0. First, we derive the statistical law (b), and then we substitute the initial condition (c) of c; having done so, we can logically deduce the probability that Ge,t_0 will happen (not that Ge,t_0 actually happened). If we want to explain why Ge,t_0 actually happened, then we need a further step: the parenthetic addendum (e).

For example, the decay of uranium-238 emits an α particle at a certain time. To explain the phenomenon, we need to calculate the probability of decay according to the half-life period of uranium-238: $1\text{-}\exp(-\lambda_{238}\times\theta)$. Then we calculate the event propensity, or the decay probability r of uranium-238 and the probability that it emits α particles at that time: $r=1\text{-}\exp(-\lambda_{238}\times\theta)$. Regardless of whether the probability is high or low, because we have already indicated the internal mechanism of the decay of uranium, we can explain the phenomenon with the parenthetic addendum.

However, according to Hempel's IS model, given the statistical law and the initial conditions, the probability that the explanans support the explanandum (uranium-238 decayed and emitted an α particle) is r. If r is close to 1, then the explanation holds, but if r is close to 0, then uranium-238 will not decay.

Railton's criticism is that the inductive inference of the IS model does not necessarily explain a low-probability event. But if the DNP model can illustrate the internal mechanism of the event (e.g., calculate the probability of uranium-238 decaying and emitting an α particle), then even a low-probability event can be explained very well.

Railton also disagrees with Hempel's thesis of epistemic relativity and the RMS of statistical explanations. For example, if 23% of uranium-238 emits 4.13 MeV α particles, while 77% of uranium-238 emits 4.18 MeV α particles, how can we then explain that uranium-238 emits a 4.13 MeV α particle at a certain moment?

Hempel's IS explanation depends on our relative knowledge; whether or not we know that a certain portion of uranium-238 will emit different α particles with different energy will inevitably influence our explanation of the event. Therefore, the IS explanation has some degree of subjectivity and is therefore relative, but Railton thinks that his DNP explanation can avoid this problem. Because we can find the internal mechanism of emitting α particles, we can accurately calculate the propensity of uranium-238 to emit 4.18 or 4.13 MeV α particles, and finally we can achieve a complete and objective explanation.

To sum up, Railton's DNP model has five features: (1) All explanations are objective; (2) to explain is to provide relevant information; (3) explanation requires not only laws, but also an account of internal mechanisms; (4) real probability explanations need probability laws, which presupposes indeterminism; and (5) probability explanations do not need the requirement of high probability.

Railton's DNP model provides the propensity interpretation of quantum mechanics: Probability represents the physical propensity of a single-chance system producing a certain result; it can be used in the single event and has causal responsibility for the natural result. We need to calculate the probability of a certain event happening by using quantum mechanics. Thus, propensity can explain the internal mechanism of an event.

Some philosophers criticize this interpretation, saying that the application of the DNP model is sorely limited and ignores many probability events of everyday life in non-quantum mechanics circumstances, such as gambling, classical thermodynamics, insurance actuary, weather forecasts, and so on. For example, flipping a coin is a non-quantum mechanics macrophenomenon. There are too many factors that may determine the result, such as flipping angle, height, wind direction, and surface hardness. We cannot explain the internal mechanism of the flipping result; therefore, we cannot explain the event "The result of flipping the coin is the tails side up."

Another critique is that the requirements of the DNP model are too high. It may make some simple explanations very complicated. For example, ice will gradually melt in warm water, and this can be easily explained by thermodynamics. But according to Railton's DNP model, we must calculate the detailed procedure of ice melting by methods of quantum mechanics, which makes a simple explanation very complicated.

Railton's response to these criticisms is that a genuine scientific explanation is very difficult to achieve. If we cannot explain the internal mechanism, then we should not pretend to have a complete explanation. Therefore, unless we can explain the internal mechanism of the result of flipping a coin, we should not pretend to have a successful explanation. Similarly, for Railton, the event of ice melting should also be explained finally by quantum mechanics.

Summary

Scientific explanation is a relatively new issue in philosophy of science and is still under discussion (Woodward 2014). Professor Hua-xia Zhang at Sun Yat-sen University points out that there are three main approaches to scientific explanation in philosophical circles (Zhang 2002, p. 32):

(1) *The epistemic approach* follows the epistemology developed by Hempel and makes further modifications. For example, van Fraassen's pragmatics of explanation and Kitcher's unificationist model fall under this approach.
(2) *The model approach* was proposed by the two philosophers Mary Hesse and Nancy Cartwright. They believe that we must construct models and use metaphors or analogies to understand and explain the world.
(3) *The ontic approach* claims that a scientific explanation must reveal the causality and internal mechanism of a given phenomenon and explicate its place in the whole picture and hierarchical structure of nature. Salmon and Railton's accounts both belong to this approach.

As the author sees it, the most important issue of scientific explanation is how to understand scientific laws. Hempel's scientific explanation models are also called covering law models, and their main feature is based on the claim that all scientific explanations – no matter whether they use the DN, IS, or DS models – must at least cover a scientific law. But what is a scientific law? Hempel attempted to propose a general form of lawlike sentences, but he finally failed to do so. Nelson Goodman also believes that there is no specific logical form of scientific laws to be found simply by analyzing counterfactual conditionals. Hence, even the central concept of scientific explanation, scientific law, is ambiguous!

As a result, many problems with scientific explanation models may be due to the very nature of scientific laws. For instance, the problem of asymmetry probably arises because scientific laws are not just mathematical formulas, but also include physical interpretations. Therefore, a scientific law itself may be asymmetric. The asymmetry of laws determines the asymmetry of explanations and the structural non-identity of explanation and prediction.

In addition, the apparent need to modify the RMS is demonstrated by that fact that we usually regard the statement "salt dissolving in water is highly probable in five minutes" as a scientific law, yet we would never think that "hexed salt dissolving in water is highly probable in five minutes" or "salt and baking soda dissolving in water is highly probable in five minutes" are scientific laws. If we were able to clarify the nature of scientific laws, then we would be able to avoid the problem of RMS modification.

In summary, our understanding of scientific explanation involves the nature of scientific laws. To understand scientific explanations, we must first work out the nature of scientific laws themselves. Scientific laws are not just mathematical formulas; they may also include physical interpretations. Moreover, the forms of scientific laws may be various; for instance, the form of biological laws can be different from the form of physical laws. So, then, are scientific laws regulative or necessary? Questions like this demand further research and analysis. Therefore, the author believes that the establishment of scientific explanation models ultimately depends on our understanding of scientific laws.

7 Theories about the growth of scientific knowledge

How does science develop? This is also an issue in philosophy of science. This chapter will discuss the main theories about the growth of scientific knowledge and study the problem of relativism brought about by Kuhn's historicism.

Logical positivists' accumulation model

Logical positivism mainly focuses on the structure of science and pays little attention to the growth of scientific knowledge. But according to the received view, most logical positivists emphasize the inductive method and the dichotomy of observation and theory, so they advocate that scientists generalize scientific theories from observation or experimental results. With the expansion of observations and experiments, scientific knowledge gradually accumulates and eventually constitutes a grand building of science.

Accordingly, logical positivists' view about the growth of science is mainly the accumulation model. According to the accumulation model, science has a clear, logical structure, the concepts of science are clear and fixed, and the scientific method is inductive. This means the growth of science is the isomorphic accumulation of scientific theories. Therefore, logical positivists believe that the rationality of science is self-evident. Although scientific theories constantly accumulate and develop, the objective of science and the evaluation criteria are fixed. All of these provide necessary and regular bases for scientific theory choice: If the goal of science is fixed, then we can compare the advantages and disadvantages of scientific theories according to the effectiveness of their realization of this aim; if the evaluation criteria of science are constant, then these evaluation criteria provide the regular basis for scientific theory choice.

Popper's theory of continuous revolutions

Logical positivists' accumulation model seems to be line with most scientists' intuition; nevertheless, it was questioned by Popper, as he rejected logical positivists' theory–observation dichotomy, instead arguing that observation itself is theory-laden because we do even not know what to observe without the guidance of theory in some way. In addition, the description of observation also needs general theoretical terms. Popper also opposed inductive methods because Hume's problem of induction has shown that there is no logical necessity from limited observations to universal theories (see Chapter 5). What is more, if we induce any theory from theory-laden observations, there may arise the fallacy of circular argumentation.

Therefore, as discussed in a previous chapter, Popper claimed that the scientific method should instead be the method of hypothesis falsification, whose logical form is as follows (T is theory and O observational statement):

$$T \rightarrow O, \neg O$$

$$\therefore \neg T$$

According to the method of hypothesis falsification, scientists boldly put forward hypotheses, and then carefully test them through observations and experiments. If a hypothesis is falsified, we will abandon it; if not, we will keep it until the next falsification. The experiment that falsifies a theory is called the "crucial experiment," such as the Michelson–Morley experiment, which is widely thought to have decisively falsified the ether-drift hypothesis (though the real story is much more complicated).

Thus, Popper's take was that the growth of science is to put forward hypotheses boldly and then refute them carefully by means of falsification. The process is thus:

Hypothesis 1——falsification——hypothesis 2——falsification——

The growth of scientific knowledge includes continuous revolutions, so Popper's view of scientific development is also called the theory of continuous revolution.[1]

Popper sometimes denoted the growth of scientific knowledge with the following form:

P_1——TT——EE——P_2

In other words, science starts from problem 1, we propose a tentative theory, and then we eliminate errors by means of critical examination, which finally gives rise to problem 2, and so on down the line.

Kuhn's historicism and relativism

In his most influential book, *The Structure of Scientific Revolutions*, Kuhn proposed the concept of a paradigm. When a scientific theory has achieved great success and attracted a majority of scientists to do research within this framework, it becomes a paradigm. Historically, phlogiston theory and oxygen theory, Aristotelian physics and Newtonian mechanics, classical mechanics and the theory of relativity can all be regarded as different paradigms.

Kuhn's view divides the growth of science into two stages: normal science and scientific revolution. In the stage of normal science, scientists are engaged in puzzle-solving research within the paradigm; in the stage of scientific revolution, which is a paradigm shift, scientists abandon the old paradigm and turn to a new paradigm. For example, if we consider the examples of paradigms above, scientific revolutions in the history of science have included the oxygen theory's replacement of the phlogiston theory, Newtonian mechanics' replacement of Aristotelian physics, and the theory of relativity's replacement of classical mechanics.

Kuhn posited that before the establishment of a scientific paradigm, there is a stage in which schools of thoughts contend for attention and in which different schools may answer the same questions via different approaches; with the great success of one school, most or all scholars will be attracted to this school, establishing it as the paradigm, and then the scientists will engage in puzzle-solving research characteristic of normal science within that paradigm. Then, with the development of conventional science, scientists may come to find that some phenomena conflict with the paradigm itself, leading to anomalies. With the accumulation of anomalies, scientists will become less and less confident in the paradigm, leading to a crisis, though it is also possible for scientists to eventually overcome anomalies and develop normal science further within the paradigm. If new theories overcome the crises, however, and then attract most scientists, they then become a new paradigm. Scientists turn to this new paradigm for normal scientific research, and a new cycle begins. Kuhn's scientific development model can be expressed in two forms:

Pre-paradigm stage (schools of thought compete)—paradigm establishment—normal science—anomaly—crisis—scientific revolution (the new paradigm is established, starting a new cycle)

Pre-scientific stage—normal science (paradigm establishment)—anomaly—crisis—scientific revolution (the new paradigm replaces the old)—new normal science

Kuhn's great contribution to philosophy of science lies not only in his introduction of historical factors, but also in his concept of incommensurability between paradigms, which leads to the problem of relativism. The term "incommensurability" originally came from ancient Greece, meaning "no common measurement." Kuhn borrowed the term to describe how there is no common foundation between two given paradigms to rationally compare their advantages and disadvantages.

The concept of incommensurability includes three aspects: (1) different scientific standards, such as what a scientific problem is; (2) the change of concepts, as even the same concept may have different meanings, such as time and space; (3) different worldviews, which either never change or radically change (Kuhn 1970, pp. 148–150). In this way, choice of scientific theory has no rational foundation, but instead depends upon historical, social, and other contingent factors, rendering scientific claims to objectivity and rationality problematic.

Martin Curd and J. A. Cover have summarized Kuhn's account of relativism into six arguments (Curd and Cover 1998, pp. 219–226):

(1) *The theory-ladenness of observation*: What scientists observe depends on what theory they accept. This means that scientists cannot observe a phenomenon contrary to their theory; scientists who accept different theories will observe different phenomena. The examination of theory by observation is just a circular argument. We cannot find neutral evidence to make a rational choice among competing theories.
(2) *Meaning variance*: The meaning of a concept also depends on theory, so after a paradigm shift, the meaning of a concept will also change. When we make rational comparisons between the two theories, we usually need logical reasoning, which must keep the meaning of a concept fixed. The meaning variance of scientific concepts makes no neutral language the foundation of rational comparison.
(3) *Problem weighting*: In Kuhn's view, the success of science lies not in the fact that it produces true observations, but in its aptitude for puzzle-solving – that is, the ability to solve scientific problems. Different paradigms pay different amounts of attention to a given problem, so scientific theory choice also depends on the weighting of certain problems. There is no fixed algorithm or rule to make a rational choice across paradigms.
(4) *Shifting standards*: After a paradigm shift, the standard and methodology of theory evaluation will also change, but there is no higher standard against which to judge the differences of these methodologies, so there is no rational way to resolve a conflict between paradigms.
(5) *Ambiguity of shared standards*: Even for shared standards, the proponents of different paradigms can give different interpretations. Therefore, even the shared standards cannot be the common basis of rational comparison.
(6) *The collective inconsistency of rules*: Theory choice involves many methodological rules – such as simplicity, fruitfulness, accuracy, etc. – but these rules may give conflicting answers as to theory choice. Therefore, even if scientists share common rules and give the same interpretation, methodological rules cannot determine theory choice from within a paradigm.

The problem of relativism raised by Kuhn has influenced various postmodern strands of thought, such that irrationalism, constructivism, and feminism can in some ways be regarded as developments and extensions of Kuhn's historicism. Kuhn's historicism also officially declared the end of logical positivism. Since then, philosophy of science has attempted to integrate historicism and logicism so as to find a middle way.

Lakatos' scientific research programmes

Hungarian philosopher Imre Lakatos set out to integrate Popper's critical rationalism and Kuhn's historicism, proposing instead the concept of scientific research programmes.

A scientific research programme is a large theoretical system that includes two parts: hard core and protective belt. The hard core consists of the most important concepts and fundamental laws of the theoretical system, such as Newton's three laws of motion and the law of universal gravitation that constitute the hard core of classical mechanics. The protective belt, then, refers mainly to the auxiliary hypotheses that surround the hard core. For instance, the number and mass of planets in the solar system belong to the protective belt of classical mechanics.

A scientific research programme has two functions: negative heuristics and positive heuristics. Negative heuristics entails adding or modifying the auxiliary hypotheses in the protective belt to protect the hard core from unfavorable experiments or observations. For example, the discovery of Neptune and Pluto prevented Newtonian mechanics from being directly falsified by astronomical observation. Positive heuristics entails actively articulating new laws and explaining new phenomena. For example, from Newton's three laws of motion, we can develop solid mechanics, fluid mechanics, aerodynamics, and so on, so that classical mechanics can continue to progress.

A scientific research programme can also be progressive or degenerating: If a scientific research programme can allow the continuous discovery of new laws and the prediction of new phenomena, then it is progressive; if a scientific research programme is constantly challenged by anomalies, and can only passively modify the protective belt to cope with problems, then it is degenerating. In Lakatos' view, experiments and observations cannot directly falsify a scientific theory, and the development of science is actually a progressive scientific research programme replacing a degenerating scientific research programme.

On the one hand, Lakatos criticized Popper's falsification as "naïve falsificationism," which does not correspond to actual scientific research. Because scientists in practice are sometimes rather cheeky, when there is a conflict between theory and observation, scientists can often put forward auxiliary hypotheses to avoid a scientific theory being falsified by observations and experiments. Or they may simply ignore these anomalies and turn to other problems altogether. In this sense, Lakatos' view agrees with historicists' analysis of the history of science.

On the other hand, Lakatos also criticized Kuhn's historicism for falling into the problem of relativism, leading him to draw the distinction between progressive and degenerating scientific research programmes. Because the development of science is a progressive research programme replacing a degenerating one, the rationality of science is guaranteed. Nevertheless, because Lakatos' notion of the scientific research programme provides no solution to the problem of incommensurability, his defense of the rationality of science is not widely recognized in philosophy of science today.

Feyerabend: Anything goes

By contrast, Austrian philosopher Paul Feyerabend fully endorsed relativism. He further stressed the concept of incommensurability, taking the irrational factors of historicism to the extreme and becoming a thorough irrationalist. He also argued against scientific method and advocated instead "methodological anarchism" with the famous slogan: "Anything goes."

From his methodological anarchism and extreme libertarianism, Feyerabend suggested that science should develop in a diverse way. Let all disciplines develop freely, such that even traditional Chinese medicine, voodoo, and witchcraft, which are not recognized by modern science at present, have the right to exist and should be treated with tolerance. Ultimately, he would argue, science can benefit from the development of these disciplines as well.

Feyerabend thus advocated a tolerant attitude toward the development of all disciplines, but his claim that there is no difference between science and myth, along with his methodological anarchism, would also abolish the normative function of philosophy of science in scientific research if taken to its logical end.

Newton-Smith's reconstruction of the rationality of science

In the book *The Rationality of Science* (1981), Oxford professor W. H. Newton-Smith defends the rationality of science from the perspective of realism. His realism upholds the following four aspects (Newton-Smith 1981):

(1) *Ontological aspect*: Whether a scientific theory is true or false depends on how the world, which is independent of us, really is.
(2) *Causal aspect*: The evidence for a theory to be true or approximately true is also the evidence of the existence of the theoretical entities that make the theory true or approximately true.
(3) *Epistemological as*pect: In principle, there can be good reasons to decide which theory among competing ones is more likely to be approximately true.
(4) *The thesis of verisimilitude*: The historically successive theories in natural science are closer and closer to truth.

Relatedly, Newton-Smith proposes temperate rationalism from the position of realism. In his opinion, the reason why science is rational is that science is real: The goal of science is to find truth. Scientific truth is not only a true statement, but it is also a truth with the ability to explain and predict. Although we cannot strictly say that any scientific theory is true, the goal of science is to increase the degree of verisimilitude. The sign of verisimilitude lies in the increasing ability of scientific prediction and explanation. Therefore, the ultimate test for science is success in observation. Science must produce new predictions and explain existing observations. Successful scientific theory includes not only success in observation, however, but also success in theory, which means it can facilitate the prediction of new theories and the explanation of existing theories. Still, success in observation is more of a priority.

According to Newton-Smith, the goal of science – to discover explanatory truths about the world – is fixed, but scientific methods can evolve continuously. The long-range success of science then forms a feedback mechanism to evaluate scientific methodology. In short, the goal of science is to pursue truth, which constitutes the reasons to evaluate scientific methods and theories. Accordingly, Newton-Smith's famous slogan is: "Realism is the truth, temperate rationalism is the way" (Newton-Smith 1981, p. 273).

Newton-Smith's moderate rationalism of realism approach is insightful, yet realism itself remains a very controversial issue in philosophy. For example, in philosophy of science, the debate between scientific realism and anti-realism is still quite intense (see Chapter 9 of this book). Even if realism is justified, moreover, we still have to face other philosophical problems. Although science is generally regarded as the model of rationality, philosophers also wish to provide reasonable defenses for other human activities, such as morality and art. If there is no truth in morality or art, is there no room for rationality? We seem to be in a dilemma: Either we provide realistic explanations for morality or art, or we must acknowledge that morality, art, and other humanities are irrational because they are not "real" in the same sense as the research objects of natural science.[2]

Laudan's non-holistic picture

In his book *Science and Value*, published in 1984, Larry Laudan criticizes Kuhn's holist picture of scientific change, and replaces the traditional hierarchical structure with the reticulated model to maintain the rationality of science. In Laudan's view, the scientific rationality model presupposed by the traditional philosophy of science is instrumental rationality. Instrumental rationality is characterized by hierarchical structure, which determines the rationality of lower-level behaviors by the effectiveness of upper-level goals (see Chapter 11). Laudan divides Kuhn's concept of paradigm into three levels: (1) ontology, which provides a conceptual framework for explaining phenomena; (2) methodology, which includes mechanical algorithms, as well as some constraints or injunctions concerning our attributes, such as seeking independent testability or avoiding ad hoc-ness; and (3), axiology, which specifies the aims of science.

These three levels also constitute the hierarchical structure of scientific rationality. Axiology is at the top, methodology in the middle, and theories about facts at the bottom. When scientists disagree with some facts or theories, they usually solve the disagreement from the perspective of methodology; if these scientists have a dispute about methodology, they must solve it from the perspective of axiology; but if scientists debate about the aims of science, they cannot find a higher level by which to judge. So, in Kuhn's view, there may be no common axiology between different paradigms, which can make a paradigm shift a radical, irrational change. The simple hierarchical model of rational consensus formation can be roughly written as follows:

Table 7.1 The simple hierarchical model of rational consensus formation (Laudan 1984, p. 26)

Level of disagreement	*Level of resolution*
Factual	Methodological
Methodological	Axiological
Axiological	None

With this noted, Laudan thinks the actual model of scientific development does not correspond to the hierarchical structure. Certainly, scientific methodology provides the basis for justification of a scientific theory, and axiology, or the aims of science, provides the basis for scientific methodology. On the other hand, however, scientific theories themselves delimit the adequate methodology, scientific methodology should also show the realizability of axiology, and, finally, there must be harmony between scientific theories and axiology. Laudan thus proposes the triadic network of justification:

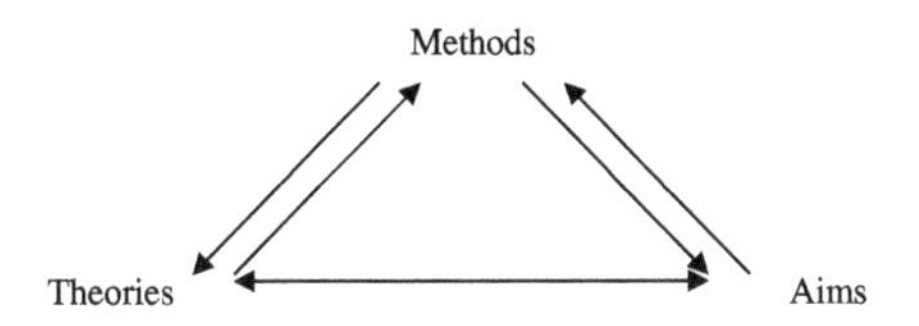

Figure 7.1 The triadic network of justification (Laudan 1984, p. 63)

Kuhn was of the opinion that a scientific revolution, or paradigm shift, is a global change, including a change of ontology, methodology, and values, and that the change is accomplished with one stroke. Kuhn's holistic model of paradigm shift can be expressed as follows:

Paradigm 1(ontology1, methodology1, values1)

↓

Paradigm 2(ontology2, methodology2, values2)

Figure 7.2 Kuhn's picture of paradigm shift

Laudan also opposes Kuhn's holistic model of paradigm shift. Laudan instead believes that actual scientific changes are completed part by part. We can thus explain the rationality of a paradigm shift by means of a non-holistic picture of scientific change and a reticulated model of scientific rationality.

For example, if a paradigm shift is due to the development of the traditional paradigm, scientists find that some part of their own paradigm, such as some theory, has problems. At this time, because other elements, methodology and values, are still unchanged, these common elements become the Archimedes fulcrum of rational comparison. The paradigm shift is not a radical change, but a part-by-part process. It may develop from T_1, M_1, A_1 (T represents theory, M methodology, A axiology or values) to T_2, M_1, A_1, then to T_2, M_2, A_1, and finally to T_2, M_2, A_2. Historians may think, from a distance, that this is a radical change from paradigm 1 (T_1, M_1, A_1) to paradigm 2 (T_2, M_2, A_2), but, in fact, only a small part of the paradigm changes at each step, and other elements are fixed, so the change is rather explained by the reticulated model of rationality.

A cross-traditional paradigm shift is also completed part by part. Scientists in paradigm 1 may initially find that, even according to their own methodology M_1 and value A_1, T_2 in paradigm 2 is still superior to T_1, so T_2 replaces T_1. However, according to A_1 and T_2, scientists working in paradigm 1 will find that M_2 is more able to realize its own values and provide explanations for the theory than M_1, so M_1 is also replaced by M_2. After the change of methodology and theories, and in order to be consistent with them, values may have to change to A_2 also. Therefore, the scientists in paradigm 1 finally give up their own tradition and accept the new tradition. Since this process is also completed in part, the change can also be explained by the reticulated model of rationality (Laudan 1984, pp. 75–76).

With that said, compared with Newton-Smith's moderate rationalism, there is no realistic presupposition in Laudan's defense of scientific rationality. Moreover, in Laudan's reticulated model, rationality is not only an instrumental tool to justify the distinction of the lower level (methodology and theory) from the upper level (values), but also a skill to achieve reflective equilibrium between values, methodology, and theory, which undoubtedly extends the traditional conception of rationality. On this account, Laudan's model of scientific development and his rational defense seem the most desirable.

Summary

Proposed models of scientific development and change range from logical positivists' accumulation model, to Karl Popper's model of continuous revolutions, and then to Kuhn's theory of paradigm (normal science plus scientific revolution). However, Kuhn's historicism, though widely accepted, also brought about the problem of relativism. Imre Lakatos thus tried to integrate Popper's critical rationalism with Kuhn's historicism, but his model of scientific research programmes failed to solve the problem of incommensurability, leading Paul Feyerabend simply to accept relativism and advocate for methodological anarchism. Newton-Smith's view, in contrast, sees the development of science as a process of approaching the truth, yet his temperate rationalism must presuppose realism, which itself remains controversial. Finally, Laudan has put forward a reticulated model of scientific rationality, advocating for a non-holistic model of paradigm shift, and this approach stands as the most desirable at present.

Science is constantly developing, and a fundamental purpose of philosophy of science is to promote the better development of science. Discussions among philosophers of science on the various possible models of scientific development have only served to deepen our understanding of science, and will undoubtedly promote the growth of science even more moving forward.

Notes

1 Popper's theory of continuous revolution seems radical, but, in fact, his view is relatively conservative, as he not only advocated criticism but also suggested that we should preserve hypotheses which are not yet falsified. Thus, in politics, Popper rejected rational construction and advocated continuous improvement of tradition. In fact, it may be more appropriate to call his view a "theory of continuous improvement" or a "theory of continuous modification."

2 Here the author does not argue against the claim that there can be realistic explanations for morality, art, and other humanities. In fact, there can be moral realism in ethics. Popper also defends the objectivity of art and value. The author simply worries that Newton-Smith's realism approach may limit the scope of rationality.

8 Demarcation between science and pseudoscience

The demarcation criterion between science and pseudoscience is another important issue in philosophy of science. In fact, in the issue-centered textbooks of philosophy of science, demarcation is often put as the first topic. So-called demarcation is sought in order to separate science from other disciplines of human beings.

In comparison, the complementary set of science can be called "non-science." The field of non-science is very broad, including religion, art, morality, and so on. To note the fundamental difference, we might observe that there is no true or false (in the scientific sense) in the field of art, in the sense that we cannot say that Ludwig van Beethoven's music is true or false, but can only evaluate its greatness according to other standards; likewise, it would make less sense to criticize William Shakespeare's *Romeo and Juliet* for not conforming to historical facts than to assess its literary value.

We should reiterate that non-science is not shameful or inferior. Indeed, Albert Einstein even felt that moral exemplars are greater than people who are engaged in scientific research, writing as much in a letter from September of 1937:

> But let us not forget that knowledge and skills alone cannot lead humanity to a happy and dignified life. Humanity has every reason to place the proclaimers of high moral standards and values above the discoverers of objective truth. What humanity owes to personalities like Buddha, Moses, and Jesus ranks for me higher than all the achievements of the enquiring and constructive mind.
>
> (Einstein 2013, p. 70)

Nevertheless, when non-science tries to pretend to be science, this is when it becomes something else: pseudoscience. The prefix "pseudo" not only points to its falsity but, more importantly, to its attempt to masquerade as science. An example in another sphere would be that a villain becomes a hypocrite when he pretends to be a gentleman; similarly, a real white paper becomes a counterfeit when someone pretends that it is in fact money.

Most non-sciences do not presume to call themselves sciences – for example, artists would rarely call themselves scientists. However, because pseudoscience pretends to be something it is not, it may bring great harm to humans and society. In recognition of this, discussions of demarcation in philosophy of science generally attend to the demarcation criterion between science and pseudoscience.

Historical background

The demarcation problem is not only a pure academic issue within philosophy of science, as it has important practical significance. For example, in the United States, where science is highly developed, the number of astrologers is still about ten times that of astronomers. Indeed, many kinds of pseudoscience can still be found if we look.

In 1981, an Arkansas state law titled "Balanced Treatment for Creation-Science and Evolution-Science Act" was passed. At that time, public schools in the United States taught Darwin's theory of evolution, but some religious people supported so-called "creation science." Creation science tries to integrate the creation theory based on belief in God with some modern scientific theories, holding, for example, that: (1) The universe was created suddenly from nothing; (2) although natural selection has played a certain role, it is insufficient to explain the origins of all creatures; (3) most animals and plants have been created and only change within a very limited range; (4) humans and apes have different ancestors, so humans are not evolved from apes; (5) macro-scale geographical changes have been caused by major disasters, such as the great flood; and (6) the Earth and living creatures have relatively recent beginnings, within the last tens of thousands of years (Curd and Cover 1998, p. 75).

Proponents of creation science thus proposed an act in Arkansas requiring teachers to give balanced treatment to both creationism and evolutionary science in public schools. This caused a large debate within educational and judicial circles centered around whether or not creation science is in fact science. If it is science, then it can be taught in public schools; however, if it is not science, but only religious belief, then according to the principle of the formal separation of church and state in modern countries, it should not be taught in public schools.

The central problem involved in this debate, then, is what comprises the features of "science." Finally, Judge William R. Overton ruled that genuine science has five essential characteristics:

> (1) it is guided by natural law; (2) it has to be explanatory by reference to natural law; (3) it is testable against the empirical world; (4) its conclusions are tentative, i.e., they are not necessarily the final word; and (5) it is falsifiable.
>
> (Curd and Cover 1998, p. 76)

Thus, Overton judged that creation theory is not based on a law of nature but on unfalsifiable dogma, meaning it is not scientific in the same sense.

Overton's court decision also attracted the attention of scientific and philosophical circles. Although many philosophers agreed with Overton's decision, they did not agree with the reasons he gave. Philosophers of science believe that there should be more detailed and pluralist reasons for demarcation. Therefore, the incident of creation science also prompted further discussion on the demarcation criterion between science and pseudoscience.

The issue of demarcation has also been of great practical significance in China. In the 1980s and 1990s, pseudosciences were widespread in China. For instance, somatic science was very popular at that time, which, as the name suggests, takes the human body as its object. Research on the human body is certainly meaningful, but under the aegis of this interest, many abnormal phenomena associated with extraordinary human abilities and Qigong[1] became the focus of attention.

For example, in 1979, a boy in Dazu County, Sichuan Province claimed to have the extraordinary ability to recognize colors by ear. The Sichuan Academy of Medical Sciences went to investigate and found that the boy cheated in the test process, but he was very skillful at this and might have received magic training. The whole incident was not reported to the public, but after that, extraordinary human abilities became very popular points of interest in the country. For example, Bao-sheng Zhang (张宝胜), a so-called "person of unusual ability," claimed to have the extraordinary abilities to read by ear and remove things from bottles without opening them. So, in 1988, Zuo-xiu He (何祚庥), an academician of the Chinese Academy of Sciences, and a magician tried to examine Bao-sheng Zhang's extraordinary abilities. When Zhang's performance did not succeed, he put the blame on some audience members' insincere intentions. Zuo-xiu He declared the failure of the experiment in public.

During the 1980s, Xin Yan (严新) was a disciple of Qigong master Haideng (海灯), who was also famous in China. In 1987, Xin Yan even cooperated with a university to study Qigong by scientific experiment, as he claimed to be able to change molecular structure with his Qigong. A fire broke out on the Greater Khingan Mountains in 1987, and Xin Yan claimed it was he who put out the fire, giving the reason that Qigong can increase both air pressure and temperature, and high air pressure can force underground water to the surface, while high temperature can make water evaporate. The accumulation of water vapor brought rain, which eventually put out the fire. In response to his claims, many people questioned why he did not put out the fire as early as possible, instead of waiting until the end of the fire. As we might expect, Xin Yan's Qigong experiments have also yielded no remarkable results.

Then there was Hongcheng Magic Liquid, another craze at that time. Hong-cheng Wang (王洪成) claimed in 1983 that he could turn regular water into fuel by simply dissolving a few drops of his magic liquid in it, and he attracted many investors and earned a lot of money as a result, though his claims were never borne out.

A final example from China at the time is related to superstitions around the *Book of Changes* (*Yijing*), a book originally used for fortune telling with milfoil or tortoise shells in ancient China. According to Chinese legend, it was written by King Wen of Zhou and later developed by Confucius himself, making it a philosophical classic of ancient dialectics; the motto of Tsinghua University – "Self-discipline, social commitment" – even comes directly from the hexagrams of this book. But in the 1980s and 1990s, many people used the book to try and predict their own fortunes and even the rise and fall of stock prices (with often rather unfortunate results) (He 1996).

As a response to once-rampant pseudosciences, 112 academicians at the Chinese Academy of Sciences jointly signed the "Initiative on the Popularization of Science" at the National Science Congress in 1995, advocating for science and opposing pseudoscience. Then, in 1999, because of the negative social influence of Hong-zhi Li's Falun Gong, China began to oppose pseudoscience on an even larger scale.

The logical criterion

Logical positivists' verification criterion

Logical positivists focus on rejecting metaphysics, so they mainly propose the meaning criteria to distinguish science from metaphysics. They tend to think that scientific propositions – i.e., meaningful propositions – satisfy the testability or translatability criteria (see Chapter 4 for details). Those which cannot satisfy the meaning criterion then belong to "meaningless" metaphysics. Likewise, their demarcation criterion, if successful, shall be deemed logical and absolute.

Herbert Feigl (1902–1988), a member of the Vienna circle, once summed up five characteristics of science, which are worth recounting. Feigl was born into a Jewish family in Reichenberg, Austria (nowadays Liberec, Czech Republic) and was initially interested in physics and chemistry, but because of his admiration for Einstein, he later turned to philosophy of science. He received his Ph.D. in philosophy, studying the relationship between chance and law in the natural sciences under the supervision of Moritz Schlick. In 1930, Feigl was funded by the Rockefeller Research Fellowship and went to Harvard University for an eight-month study. From 1931, he taught at the University of Iowa and became a full professor at the University of Minnesota from 1940 until his retirement in 1971. His major works include *The Mental and the Physical* (1967) and *Inquiries and Provocation* (1980).

In Feigl's view, the five criteria of science are (Klemke et al. 1998, p. 32):

(1) *Intersubjective testability*: Logical positivists do not like the word "objectivity" so much, as it is thought to go beyond the scope of our experience and it has also been challenged by quantum mechanics. They instead

prefer the term "intersubjectivity," which requires that scientific theories can be tested publicly and that experimental results can be repeated.

(2) *Reliability*: Theory is not only testable, but is also true after testing.

(3) *Definiteness and precision*: Scientific concepts must be definite and laws of nature precise.

(4) *Coherence or systematic character*: The whole theoretical system is not merely a bunch of true statements about nature, but a hierarchical structure in which scientific theories are related to and consistent with each other.

(5) *Comprehensiveness or scope*: The theoretical system is complete and can explain natural phenomena to the greatest extent.

According to Feigl, genuine sciences satisfy all five of these characteristics, but pseudosciences like astrology do not. Astrology is a long-standing theory in the West that claims to predict someone's fortune according to the constellation under which he/she was born. Astrology originated very early and established a full-fledged system around 700 BC. Brought to Greece in the wake of Alexander the Great's expeditions, it later gained influence in the Roman Empire, was further developed by Claudius Ptolemy, and reached its peak in the second century AD.

However, if we examine astrology carefully, it might also satisfy Feigl's scientific criteria. First, some predictions of astrology are testable. For instance, Michel Gauquelin published a book on astrology in 1967 in which he surveyed 25,000 French people and concluded that the fate of human beings has little to do with the positions of the Sun and the Moon at birth, but does appear to be more related to the positions of planets to some extent. Second, astrology is also sometimes effective, perhaps explaining why there are still many people who believe in it. Third, astrology also entails precise observations of constellations. Fourth, astrology can form a system of its own, and finally, astrology is used to explain many phenomena. In this way, Feigl's demarcation criteria cannot entirely exclude astrology from science.

Popper's falsificationism

Karl Popper, born into a Viennese Jewish family in 1902, was deeply influenced (though not always positively) by three great Jewish thinkers: Karl Marx, Sigmund Freud, and Albert Einstein. In the years of his youth, he upheld Marxism for a time and joined the Association of Socialist School Students. Popper's father also knew Freud's sister well, and Popper served briefly as a voluntary social worker in a psychiatric clinic of Alfred Adler, a student of Freud. Popper was a good friend of Einstein and offered a philosophical interpretation for his theory of relativity. Let us consider how he viewed these thinkers in turn.

At that time, Freud's psychoanalysis, which used sexual impulses to explain all human behaviors, was all the rage in Vienna. During the first half of the twentieth century, the social climate in Europe was still very conservative, only becoming more open after the sexual liberation movement in the 1960s and 1970s, so it may very well be that many psychological problems in the early twentieth century were caused by sexual repression. Nevertheless, Freud extended his research results to an excessive extent, thinking that all human behaviors confirmed his sexual psychology. If someone denied it, it could simply be claimed that he/she was too sexually repressed to acknowledge the truth.

Similarly, Freud's student Alfred Adler created an individual psychology, using the inferiority complex to explain all human behaviors: All are caused by, or aim to overcome, an inferiority complex. Despite the insights raised, he also extended his individual psychology too far by explaining all human behaviors with the inferiority complex. If someone denied it, though, this was simply because he/she was afraid or unwilling to admit the truth due to his/her inferiority complex.

Both Freud's and Adler's academic styles were overly dogmatic, regarding all cases of human behavior as confirmation of their own theories, such that in no case could their theories be falsified. For example, if someone were to jump into the water and rescue a person, the Freudian school could claim that this is the result of overcoming sexual repression, and the Adlerian school could say it is the result of surpassing inferiority. On the other hand, if the person does not jump to the rescue, the Freudian school could say that this is the result of sexual repression, and the Adlerian school could blame it on an inferiority complex. Thus, no matter what the result is, it will always serve as confirmation for the given theory and cannot be falsified.

Popper was disgusted with this sort of dogmatic thinking. He was also against dogmatic Marxism and published two books criticizing it: *The Open Society and Its Enemy* (1945) and *The Poverty of Historicism* (1957). For example, in his book *The Poverty of Historicism*, he sums up Marx's historical materialism as a kind of "historicism" and then raises the following criticisms (Popper 2013, pp. xi–xii):

(1) The course of human history is strongly influenced by the growth of human knowledge.
(2) We cannot predict, by rational or scientific methods, the future growth of our scientific knowledge.
(3) We cannot, therefore, predict the future course of human history.
(4) That means we must reject the possibility of a theoretical history; that is to say, of historical social science that would correspond to theoretical physics.
(5) The fundamental aim of historicist methods is therefore misconceived, and historicism collapses.

Popper claimed 1919 was a decisive year for him, as it was that year that the bending of stellar light predicted by Einstein's theory of relativity was confirmed by the British astronomer Sir Arthur Eddington's observations in West Africa. Before that, Einstein had said that his theory was open to the challenge of observation and that if resulting observations were inconsistent with his own theory, he would finally give it up. Popper deeply admired Einstein's attitude of welcoming criticism and came to believe that this was the properly scientific attitude. Thus, by the end of 1919, Popper had come to the conclusion that the scientific attitude was a critical one (Popper 1974, p. 29).

Due to this background, Popper focused on the question "When should a theory be ranked as scientific?" or "Is there a criterion for the scientific character or status of a theory?" He also criticized logical positivists for confusing the three concepts of verifiability, meaningfulness, and scientific character. In his view, the problem of meaningfulness is a pseudo-problem, for science cannot completely exclude metaphysics. As a result, he proposed the concept of falsifiability as the demarcation criterion between science and pseudoscience, rather than the meaning criterion between science and metaphysics. His claims to this point are thus (Popper 1962, pp. 36–37):

(1) It is easy to obtain confirmations, or verifications, for nearly every theory—if we look for confirmations.
(2) Confirmations should count only if they are the result of risky predictions; in other words, if, unenlightened by the theory in question, we should have expected an event which was incompatible with the theory—an event which would have refuted the theory.
(3) Every 'good' scientific theory is a prohibition: it forbids certain things from happening. The more a theory forbids, the better it is.
(4) A theory which is not refutable by any conceivable event is non-scientific. Irrefutability is not a virtue of a theory (as people often think) but a vice.
(5) Every genuine test of a theory is an attempt to falsify it, or refute it. Testability is falsifiability; but there are degrees of testability: some theories are more testable, more exposed to refutation, than others; they take, as it were, greater risks.
(6) Confirming evidence should not count except when it is the result of a genuine test of the theory; and this means that it can be presented as a serious but unsuccessful attempt to falsify the theory.
(7) Some genuinely testable theories, when found to be false, are still upheld by their admirers—for example by introducing *ad hoc* some auxiliary assumption, or by reinterpreting the theory *ad hoc* in such a way that it escapes refutation. Such a procedure is always possible, but it rescues the theory from refutation only at the price of destroying, or at least lowering, its scientific status.

One can sum up all this by saying that the criterion of the scientific status of a theory is its falsifiability, or refutability, or testability.

There are also degrees of falsifiability: The more content a statement contains, the more likely it is to be falsified. For example, "It will rain tomorrow" will be falsified by the fact that it did not rain the next day. "It will rain tomorrow afternoon" has more content and is more likely to be falsified. If it did rain, but only in the morning, it will also be falsified. By the same token, "It will rain 20 mm from 2:00 to 5:00 tomorrow afternoon" contains more content and can be falsified to an even higher degree.

Therefore, Popper regards falsifiability as the demarcation criterion between science and pseudoscience. The higher the degree of falsifiability, the more scientific it is. Because the falsifiability criterion uses the method of hypothesis falsification, it is logical and absolute.

Still, Popper's falsifiability criterion also has its difficulties. First of all, although some propositions meet the falsifiability criterion, we never regard them as science. For example, "9999 people who are hit on the head with a baseball bat can become a fairy" has a high degree of falsifiability, but we will not regard it as a scientific hypothesis. Therefore, falsifiability can only be regarded as a necessary condition, not a sufficient condition, of the demarcation criterion.

In addition, there are many cases in the history of science that violate the falsifiability criterion. For example, the discoveries of Neptune and Pluto were the result of scientists' refusal to falsify. These are problems that Popper's naïve falsificationism simply cannot answer.

Relative criterion

Kuhn's blurred criterion

Kuhn's concepts of paradigm and incommensurability make the demarcation more complicated. In his view, the development of science is a process of normal science (i.e., scientific research within the paradigm) and scientific revolution (i.e., paradigm shift), but different paradigms (normal sciences) are incommensurable, without a common basis for rational comparison. For that reason, modern science and ancient thought are merely different paradigms, without any fundamental distinction between science and pseudoscience.

As mentioned, Popper regarded astrology as a typical pseudoscience for two reasons: (1) Astrologers refuse to refute their theory, and (2) the explanations and predictions of astrology are ambiguous. Kuhn, however, argued against Popper in his paper "Logic of Discovery or Psychology of Research?" First of all, he maintained, astrology has recorded many prediction failures in history, and even astrologers admit that such failures happen frequently. Secondly, astrologers think that predicting the future is extremely complicated; it involves the precise position of the astronomical constellations and the exact birth time of some given person, making it easy to make mistakes. Nevertheless, exact sciences such as physics and astronomy may also resort to the same defense, so astrology's explanation of failures is not necessarily unscientific. It is only because astrology is no longer trustable that its arguments become fallacious (Kuhn 1970, pp. 1–23).

Of course, Kuhn's position fluctuated at times: On the one hand, he put forward the relativity of science; on the other hand, he was unwilling to give up the rationality and objectivity of science. Thus, his work blurred the demarcation between science and pseudoscience.

Lakatos' demarcation criterion

Lakatos believed that the demarcation between science and pseudoscience is not only a philosophical issue, but it also has significant social and political relevance. For example, the Catholic Church banned Copernicus' theory in 1616, and many Copernican scholars were judged; more recently, in 1949, the Soviet Union declared Mendelian genetics to be pseudoscience and sent many geneticists to concentration camps.

Lakatos criticized traditional philosophers of regarding whatever hypothesis everyone believed as science. But, in fact, scientists tend to be skeptical about even the most advanced scientific theories, whereas only fanatical believers trust in doctrines without doubt. Therefore, he charged, "Blind commitment to a theory is not an intellectual virtue: it is an intellectual crime" (Lakatos 1978, p. 1).

Lakatos also rejected the logical empiricist criterion of demarcation because scientific theory, as well as other hypotheses, is impossible to verify. Though supporters of inductive logic maintain that scientific theories have a higher degree of confirmation, Lakatos believed, following Popper, that scientific theories and other hypotheses are equally "improbable."

Lakatos argued that Popper's demarcation is not the criterion of distinguishing between scientific theory and pseudoscientific theory, but between a scientific attitude and a pseudoscientific attitude. For example, if Freudian scholars are willing to give up their theory when they observe some counterexample, their attitude is scientific; otherwise, it is pseudoscientific. Nevertheless, Lakatos criticized Popper's falsificationism as naïve for not corresponding to actual scientific practice. Because scientists are sometimes shameless, when a theory collides with observations, they either introduce auxiliary hypotheses to explain an anomalous phenomenon, or they simply ignore the anomaly and turn to other problems.

According to Lakatos, the unit of demarcation should be scientific research programmes, not individual hypotheses. As discussed above, a scientific research programme consists of a hard core and a protective belt. The hard core refers to the core theories and concepts, such as the three laws of Newtonian mechanics and the law of universal gravitation in classical mechanics. The protective belt mainly refers to the auxiliary hypotheses surrounding the hard core. The protective belt includes both negative and positive heuristics, which promote the development of scientific research programmes.

Lakatos thought there is no such thing as a "critical experiment," and scientific hypotheses are not directly falsified by observation or experiments. Scientific tests have a certain tenacity, thus requiring long-term and global examination. In this way, scientific progress is actually a progressive research programme replacing a degenerating research programme. Therefore, Lakatos reduced the issue of demarcation criterion to the issue of scientific evaluation: How we can distinguish progressive programmes from degenerating programmes.

Progressive scientific research programmes can constantly respond to the challenges of anomalies and predict new facts, while degenerating programmes can only passively cope with anomalies and rarely predict new facts. However, the concepts of "progressive" and "degenerating" no longer constitute a logical and absolute criterion, but rather involve the subjective judgment of human beings and have a historical dimension. For example, the yin-yang theory in China was a progressive research programme two thousand years ago, but now it is degenerating. Lakatos' demarcation criterion thus retains some historicity and relativity.

But Lakatos still maintained that the key to distinguishing progressive programmes and degenerating programmes is not whether they can be confirmed or falsified, but whether they have produced successful predictions. For example, in 1705, Edmond Halley predicted the return of Halley's comet in 1758 with the use of Newtonian mechanics, and his prediction became true 53 years later. Einstein's general theory of relativity also predicted that the light coming from the stars was bent as it passed the Sun, and this was confirmed by astronomical observations in 1919. Therefore, a distinction based on Lakatos' demarcation criterion is logical and empirical, having some degree of absoluteness (Lakatos 1978, pp. 1–7).

On account of these movements mentioned above, Chinese scholar Chen Jian argued in his Ph.D. dissertation, which was later published as a monograph, that from logical empiricism and Popper's naïve falsificationism to Kuhn's historicism and Lakatos' falsificationism, "we see a chain of criticism. The first two segments of the chain are absolute criteria, and the last two are Kuhn's softening and Lakatos's tenacity" (Chen Jian 1997, pp. 43–44).

Dissolving demarcation criteria

Paul Feyerabend followed Kuhn's historicism, but went further in developing his epistemological anarchism, arguing that there is no such thing as the scientific method and the only principle in science is "Anything goes."

Feyerabend's masterpiece is *Against Method* (1975). He said he was short of funds at the time, so when he was invited to write a book on the relation between science and religion, in order to make the book sell, he had to make it provocative. His most provocative statement about the relation between science and religion was that "science is a religion" (Feyerabend 1998, p. 54). Though initially posed to be provocative, he soon realized, however, that such a view was not unreasonable. So, gradually, he came to the personal conclusion that "anarchism, while perhaps not the most attractive political philosophy, is certainly excellent medicine for epistemology, and for the philosophy of science" (Feyerabend 1975, p. 17).

Feyerabend had two reasons for his epistemological anarchism: (1) The world, which we want to explore, is a largely unknown entity; therefore, we must keep our options open and restrict ourselves in advance. (2) Scientific education cannot be reconciled with a humanitarian attitude; it is in conflict "with the cultivation of individuality which alone produces, or can produce, well-developed human beings" (Feyerabend 1975, p. 20).

According to Feyerabend, the history of science and technology tells us that any method, no matter how reasonable or based on epistemology, will one day be violated. For instance, ancient atomism, the Copernican Revolution, modern atomism, and the wave theory of light are all due to the fact that scientists intentionally or unintentionally got rid of the inherent epistemological rules. He suggested, as a result, that we should be against method or rule.

Taking the "consistency condition" as an example, Feyerabend argued that (1) the evidence for rejecting a theory normally arises from another incompatible theory. The development of human knowledge is not a progressive approximation to the truth, but an ever-increasing ocean of mutually incompatible hypotheses. This set of alternatives is formed by theories and fairy-tales alike, and in the process of their competition, human consciousness is developed. (2) Since observation is theory-laden, there can be a suspicion of circular argumentation between theories and facts. The problem cannot be solved from the inside; hence, an external standard of criticism is needed. Inconsistency with facts is therefore desirable.

In fact, science is always and everywhere enriching itself with unscientific methods and unscientific achievements. For example, astronomy has learned a great deal from Pythagoreanism and Platonism, while modern medicine may also benefit in certain ways from herbal medicine, witchcraft, midwives, and drug dealers. He thus advocated that the demarcation between science and non-science is not only artificial, but also harmful to the development of knowledge. In order to understand nature and control the environment, we must use all ideas and methods. He wrote:

> The examples of Copernicus, the atomic theory, voodoo, Chinese medicine show that even the most advanced and the apparently most secure theory is not safe, that it can be modified or entirely overthrown with the help of views which the conceit of ignorance has already put into the dustbin of history. This is how the knowledge of today may become the fairy-tale of tomorrow, and how the most laughable myth may eventually turn into the most solid piece of science.
>
> (Feyerabend 1975, p. 52)

Feyerabend also charged that demarcation can be morally harmful from the perspective of a free society and humanitarianism. He pointed out that in the seventeenth and eighteenth centuries, science, as an ideological weapon against feudalism and religion, was an instrument of liberation and enlightenment. In fact, China also used Mr. Science and Mr. Democracy as liberating forces against imperialism and feudalism in the May Fourth Movement of 1919. Yet there is nothing inherent in science that makes it essentially liberating. In fact, because modern science only teaches "facts" and has lost its former critical attitude, it has become a sort of scientific chauvinism that oppresses other forms of thought.

Feyerabend claimed that diversity of opinion is necessary for objective knowledge, and to encourage diversity is compatible with humanitarianism because any condition of consistency will limit diversity. Science pursues truth, but Feyerabend maintained that truth is not the only value, as there are also such values as freedom and independent thinking. On this account, he criticized modern science for "imprisoning freedom of thought," writing:

> Thus Science is much closer to myth than a scientific philosophy is prepared to admit. It is one of the many forms of thought that have been developed by man, and not necessarily the best. It is conspicuous, noisy, and impudent, but it is inherently superior only for those who have already decided in favour of a certain ideology, or who have accepted it without ever having examined its advantages and its limits. And as the accepting and rejecting of ideologies should be left to the individual it follows that the separation of church and state must be complemented by the separation of state and science, that most recent, most aggressive, and most dogmatic religious institution. Such a separation may be our only chance to achieve a humanity we are capable of, but have never fully realized.
>
> (Feyerabend 1975, p. 295)

Feyerabend proposed abandoning the demarcation between science and pseudoscience from both aspects of knowledge and morality. His criticism was reasonable in the context of opposing the chauvinism of science and, properly considered, can serve to make science more tolerant. Nevertheless, his dissolution of a demarcation criterion is helpless for counteracting pseudosciences that can, in many cases, do great harm to human life and society. These harms cannot be dissolved, but can only be overcome. For this reason, philosophers have more recently put forward a pluralist criterion of demarcation.

Pluralist criterion

Thagard's three elements criterion

Paul R. Thagard, a Canadian philosopher of science at the University of Waterloo, has raised three metaphilosophical questions about demarcation: "(1) Why is it important to demarcate science and from what should it be distinguished? (2) What is the logical form of a demarcation criterion?

(3) What are the units that are marked as scientific or pseudoscientific?" (Thagard 1988, p. 157).

Traditional demarcation criteria regard propositions or theories as the units of demarcation: Are certain propositions or theories scientific or pseudoscientific? Thagard posits that the demarcated objects should be seen as fields, which can be understood as historical entities embracing theories, their applications, and practitioners. This is not only an epistemic notion, but it also has social and historical dimensions, providing the possibility for a pluralist criterion of demarcation.

Traditionally, the ideal form of the demarcation criterion was thus: X is scientific if and only if C, where X is a theory, proposition, or field, and C are necessary and sufficient conditions for X being scientific. Thagard thinks this form of definition is quixotic and can never succeed. He proposes a list of features that are typical of science and a contrasting list of features that are typical of pseudoscience. To judge whether X is scientific or a pseudoscientific, then, is to see whether its features are closer to what is typical of science or of pseudoscience. His list of features of typical science and pseudoscience are as follows:

Among the above profiles, thinking by correlation infers that "two things or events are causally related from the fact that they are correlated with each other"; while resemblance thinking infers that "two things or events are causally related from the fact that they are similar to each other." Modern sciences use correlative thinking, while traditional witchcraft mainly appeals to resemblance thinking. For example, ancient astrologers associated the red appearance of Mars with war, so they called the planet "Mars" after the Roman god of war; the beautiful Venus was associated with beauty, motherhood, birth, and destiny, so the planet was called "Venus" after the Roman goddess of love, beauty, and fertility.

Thagard adds three notes to his demarcation profiles: (1) These features are not strictly necessary or sufficient conditions of science or pseudoscience; (2) we can judge whether a field is scientific or pseudoscientific by observing whether it is closer to the left or right side; (3) conceptual profiles would be more exact by turning the features into rules with associated strength (Thagard 1988, p. 170).

Table 8.1 Thagard's profile of science and pseudoscience (Thagard 1988, p. 170)

Science	*Pseudoscience*
Uses correlative thinking.	Uses resemblance thinking.
Seeks empirical confirmations and disconfirmations.	Neglects empirical matters.
Practitioners care about evaluating theories in relation to alternative theories.	Practitioners are oblivious to alternative theories.
Uses highly consilient and simple theories.	Uses non-simple theories: many ad hoc hypotheses.
Progresses over time: develops new theories that explain new facts.	Stagnant in doctrine and applications.

In his 1978 paper "Why Astrology is a Pseudoscience," Thagard introduces a matrix of three elements for the demarcation criterion: theory, community, and historical context. The traditional demarcation criteria focus on theories, discussing whether a theory is scientific or pseudoscientific. Thagard, however, extends the framework and analyzes the community and the historical context of the theory.

Subsequently, Thagard raises three questions about the community: (1) whether practitioners are in agreement on the theory and on how to solve problems; (2) whether they care about explaining anomalies and comparing with other theories; and (3) whether they actively attempt to confirm or disconfirm their theory. In his view, the practitioners of a scientific community have a consensus and often adopt the same approach to solve problems; a scientific community is very concerned with the success of other theories; and scientists also actively attempt to confirm or disconfirm their own theories. The attitude of a pseudoscientific community is precisely the opposite. The historical context illustrates that a theory is rejected only when it has faced anomalies over a long period and has been challenged by another theory that better accounts for them.

Thus, Thagard defines a theory or discipline as pseudoscientific if and only if "(1) it has been less progressive than alternative theories over a long period of time, and faces many unsolved problems, but (2) the community of practitioners makes little attempt to develop the theory towards solutions of the problems." Astrology is therefore a pseudoscience because (1) it is unprogressive and has developed little since the time of Ptolemy, (2) problems such as the precession of equinoxes are still unsolvable, (3) there are alternative and better theories to explain personality and human behaviors (e.g., psychology), and (4) the community of astrologers is unconcerned with advancing astrology (Thagard 1978, pp. 223–234).

Of course, Thagard's demarcation criterion also brings with it some problems: (1) If there is no competing theory, can we regard the only theory – such as pyramidology, which is quite popular now – as scientific? (2) If a discipline is scientific now, is it possible for it to become pseudoscientific in the future? (3) Sometimes it would be irrational to reject a theory too early. For example, if the development of psychology only came in the late nineteenth century, should we accept astrology before that time?

Bunge's ten-tuple criterion

Mario Bunge (1919–2020), an Argentina-born philosopher of science at McGill University, has also proposed a pluralist criterion of demarcation with ten aspects in any cognitive field (Bunge 1984, p. 37):

E=(C, S, G, D, F, B, P, K, A, M)
C=Cognitive community.
S=Society hosting C.

G=General outlook, or worldview, and philosophy of the C.
D=Domain or universe of discourse of E: the objects E is about.
F=Formal background: logical and mathematical tools employable in E.
B=Specific background, or set of presuppositions about D borrowed from fields of knowledge other than E.
P=Problematics, or set of problems E may handle.
K=Specific fund of knowledge accumulated by E.
A=Aims or goals of the C in cultivating E.
M=Methodics or collection of methods utilizable in E.

Any typical science, such as physics and sociology, is a cognitive field E=(C, S, G, D, F, B, P, K, A, M) that satisfies the following conditions: (1) Each of the ten components of E changes as a result of inquiry. (2) The members of the research community C have received specialized training, hold strong information links among themselves, and initiate or continue a tradition of inquiry. (3) The society S encourages or tolerates the activities of the components of C. (4) The domain D is composed exclusively of real entities. (5) The general outlook or philosophical background G is composed of: lawfully changing concrete things; a realist theory of knowledge; a value system enshrining clarity, exactness, depth, consistency, and truth; and the ethos of the free search for truth. (6) The formal background F is a collection of up-to-date logical and mathematical theories. (7) The specific background B is a collection of up-to-date and reasonably well-confirmed data, hypotheses, and theories. (8) The problematics P consist exclusively of cognitive problems concerning nature. (9) The fund of knowledge K is a collection of up-to-date and testable theories, hypotheses, and data compatible with those of B and obtained in E at previous times. (10) The aims A include discovering or using the laws in D, systematizing hypotheses about D, and refining methods in M. (11) The methodics M contains exclusively scrutable and justifiable procedures. (12) E is a component of a wider cognitive field – that is, E overlaps with other cognitive fields. (Bunge 1984, pp. 38–39)

A field of knowledge that satisfies all the following conditions will be regarded as pseudoscience. The main features of pseudoscience include: (1) The ten components of E change little over the course of time, as a result of controversy or external pressures, rather than of scientific research. (2) The believers of the community C call themselves scientists, but they conduct very little scientific research. (3) The host society S supports C for practical reasons or tolerates C, while relegating it beyond the border of its official culture. (4) The domain D teems with unreal or at least not certifiably real entities. (5) The general outlook G includes: an ontology countenancing immaterial entities or processes; an epistemology making room for arguments from authority or for paranormal modes of cognition; a value system that does not enshrine clarity, exactness, depth, consistency, or truth; and an ethos that, far from facilitating the free search for truth, recommends the staunch defense of dogma. (6) The formal background F is usually modest: Logic and mathematical modeling are not always respected. (7) The specific background B is small or nil: A pseudoscience learns little or nothing from other cognitive fields. (8) The problematics P includes many more practical problems concerning human life than cognitive problems. (9) The fund of knowledge K is practically stagnant and contains numerous untestable or even false hypotheses. (10) The aims A of the members of C are often practical rather than cognitive. (11) The methodics M are neither checkable by alternative procedures nor justifiable by well-confirmed theories. Criticism is not welcomed. (12) It is practically isolated: There is no system paralleling the genuine sciences (Bunge 1984, pp. 39–41).

Bunge attaches great importance to anti-pseudoscience and regards the demarcation criterion as one of the central issues in philosophy of science. He writes:

> First, superstition, pseudoscience, and antiscience are not rubbish that can be recycled into something useful; they are intellectual viruses that can attack anybody, layman or scientist, to the point of sickening an entire culture and turning it against scientific research. Second, the emergence and diffusion of superstition, pseudoscience, and antiscience are important psychosocial phenomena worth being investigated scientifically and perhaps even used as indicators of the state of health of a culture. Third, pseudoscience and antiscience are good test cases for any philosophy of science.
>
> (Bunge 1984, p. 46)

With that said, Bunge's own demarcation criterion is not without its own problems. He mentions the concepts of "real entity" and "truth" in the demarcation criterion, but these concepts are still controversial in philosophy. In addition, there may be counterexamples to his demarcation criterion, such as how papers about alchemy may be deliberately written to be profound and incomprehensible, whereas papers of science popularization may aim for understanding by the general public. Before modern science, there was relatively little quantitative research in ancient sciences. Should these be classified as pseudoscience, then?

Summary

The demarcation criteria between science and pseudoscience, from those of logical positivism and Popper's logical criteria, to Kuhn's and Lakatos' relative criteria, to Feyerabend's dissolving criterion, to the pluralist criteria, have followed a very clear, internal path of development.

Since the rise of historicism, philosophers of science have generally acknowledged that there is no absolute demarcation criterion between science and pseudoscience. But the fact that there is no *absolute* demarcation criterion does not mean there is no demarcation criterion whatsoever. Some relativists or irrationalists who thus think that "Anything goes" in fact confuse these two conceptions. The new development of a pluralist criterion aims to find a historically adequate criterion by analyzing concrete cases in specific historical and social contexts.

Note

1 "Qigong can be described as a mind-body-spirit practice that improves one's mental and physical health by integrating posture, movement, breathing techniques, self-massage, sound, and focused intent" (www.nqa.org/what-is-qigong-).

9 Scientific realism

Introduction

In philosophy, there is a long-standing dispute between realism and anti-realism. Realism holds that things exist independently of our minds and cognition, while anti-realism denies this. The debate around scientific realism focuses on the question of whether theoretical terms or unobservable objects really exist.

Here we need to clarify two concepts: "theoretical term" is a linguistic concept, which is opposed to "observational term"; and "unobservable object" is an ontological concept, which is opposed to "observable object." Roughly speaking, the two concepts have the same extension. For example, in a microscopic area, the concepts of atoms, electrons, and quarks are all theoretical terms, and the objects they refer to are also unobservable. There may also be theoretical terms or unobservable objects in the macroscopic area. For example, the theoretical physicist Stephen Hawking of Cambridge University describes the concept of "black hole," which is also a theoretical term, in his famous book *A Brief History of Time*. Because the gravity of a black hole is so immense, it will attract any objects nearby, even light, so that it cannot be observed directly and belongs to an unobservable object.

According to the dichotomy of observable–unobservable objects and observational–theoretical terms, there are four main accounts of theoretical terms in philosophy of science: (1) The strongest account claims that theoretical terms are not essential and should be eliminated altogether. For example, Frank Ramsey has proposed the concept of "Ramsey sentences" to eliminate such theoretical terms. (2) A weaker version, embraced by some logical empiricists, holds that theoretical terms are ontologically meaningless but have instrumental roles in science – that is, they are instrumentally useful. (3) Operationalists suggest theoretical terms can be precisely defined by observational terms, although operationalism will encounter the problem of Carnap's reduction sentence. (4) The weakest version of the empiricist account holds that theoretical terms can be indirectly connected with observational terms, so they are partially meaningful. In summary, philosophy of science must face the problem of how to deal with theoretical terms. According to scientific realists, theoretical terms in science are as real as observational terms, while anti-realists only recognize the instrumental usefulness of theoretical terms, but deny their existence (Curd and Cover 1998, p. 1228).

There are many formulations of scientific realism. In its simplest form, scientific realism claims that science gives us the true picture of the world, faithful in its details; the entities postulated in science really exist; the advances of science are discoveries rather than inventions. Therefore, scientific realism entails two claims: A scientific theory tells us a true story about the world, and scientific research is discovery as opposed to invention.

Regarding scientific realism, Wilfrid Sellars said: "to have good reasons for holding a theory is *ipso facto* to have good reasons for holding that the entities posited by the theory exist" (Sellars 1962, p. 97). Brian Ellis' version is stronger: "I understand scientific realism to be the view that theoretical statements of science are, or purport to be, true generalized descriptions of reality" (Ellis 1979, p. 28). Hilary Putnam, influenced by Michael Dummett, gives another definition: "A realist...holds that (1) the sentences of that theory are true or false; and (2) that what makes them true or false is something external—that is to say, it is not...our sense data, actual or potential, or the structure of our minds, or our language, etc." (Putnam 1975, p. 69f). Later, Putnam also learned from Richard Boyd and claimed that "the theories accepted in a mature science are typically approximately true...the same term can refer to the same thing even when it occurs in different theories" (Putnam 1975, p. 73).

As for anti-realism, Harold I. Brown distinguishes three types of arguments: historical induction, arguments from underdetermination, and arguments from incommensurability. Historical induction draws evidence from the history of science – since so many scientific claims that were once regarded as irrefutable truth have finally been found to be false – as to how we can have adequate grounds for asserting the truth of current scientific theories. Kuhn's concepts of paradigm and incommensurability are also frequently used to argue against realism. According to conceptual relativism, what is real is relative to a paradigm. Proponents of different paradigms seem to live in different worlds, as they have very different accounts of reality. How, then, can we decide whether a scientific theory is true outside its paradigm (Brown 1990, pp. 211–242)? This is what is at stake. In what follows, we will thus introduce the main debates between scientific realism and anti-realism.

Historical clues

The historical pioneer of scientific realism is atomic realism. Since the publication of Antoine-Laurent de Lavoisier's *Traité élémentaire de chimie* (*Elementary Treatise on Chemistry*) in 1789, and especially after John Dalton proposed the atomic theory in the early nineteenth century, the "atom" has become a central concept in modern science. However, the understanding of the concept has been rife with controversy. Realism about atoms holds that atoms are as real as tables and chairs, while anti-realism denies this.

At the end of the nineteenth century, Ernst Mach, Pierre Duhem, and Jules Henri Poincaré, respectively, proposed phenomenalism, instrumentalism, and conventionalism with respect to theoretical terms such as "atom." Pierre Duhem (1861–1916) was a French physicist, philosopher, and historian of science. In his book *To Save the Phenomena*, he put forward an instrumentalist view, while in *The Aim and Structure of Physical Theory*, he developed a form of conventionalism. Poincaré (1854–1912), a great French mathematician, physicist, and philosopher, likewise proposed a kind of conventionalism, but he is also regarded as a realist because he contended that since science is based on the unity and simplicity of nature, the aim of science is constantly to pursue more general regularities.

Among these views, phenomenalism holds that physical objects can be reduced to sensory experience, and the descriptions of physical objects can be analyzed as the phenomenal statements of sensory experience. For example, the description "is white" of the physical object "snow" can be reduced to the phenomenal language "has the feeling of white." Second, instrumentalism insists that a scientific theory is not a true description of the unobservable objects, but a purely useful tool that enables us to settle down and explain the observable world. Following the underdetermination thesis, instrumentalists think, in principle, that infinite theories corresponding to our limited observation experience can be put forward, and we finally must choose theories based on their effectiveness. Third, conventionalism claims that, in principle, scientific theories cannot be directly verified or falsified. Theories, at a deep level, are selected from many possible theories, and theory choice depends on our conventions. For example, both Newtonian mechanics and quantum mechanics can explain the world. We choose quantum mechanics not because quantum mechanics is right and Newtonian mechanics is wrong, for we can neither verify that quantum mechanics is right nor that Newtonian mechanics is wrong, which in principle can avoid falsification by increasing auxiliary hypotheses. Therefore, the theory choice is primarily the result of scientists' convention.

From the theory of economy of thought, Mach also thought that unobservable objects are just convenient fictions that can be used to simplify our theories. He took an anti-realist position regarding atoms, force, and field because these concepts cannot be reduced to demonstrable sensory experience. In scientific circles, moreover, Einstein was initially deeply influenced by Mach and was once a positivist. In particular, Einstein highly appreciated Mach's criticism of Newton's concept of absolute space-time and put forward operational definitions of time and space. Nevertheless, in the 1920s, Einstein returned to realism.

With the development of quantum mechanics, many physicists, especially those accepting the Copenhagen interpretation of quantum mechanics, began to prefer anti-realism. For example, Werner Heisenberg rejected the reference of unobservable terms and realistic images of quantum mechanics. Erwin Schrödinger initially thought that his equation presupposed the existence of a "wave," but he soon gave up the realistic interpretation of Schrödinger's equation.

Niels Bohr's complementarity concept, Heisenberg's uncertainty principle, and Max Born's statistical interpretation of Schrödinger's wave function all tend to take an instrumentalist stance on science. Bohr even believed that Einstein's realism would hinder the progress of science. The anti-realist position was formally established at the Solvay Conference in October 1927, and Bohr's interpretation of quantum mechanics, called the "Copenhagen program," has become the dominant view in the community of quantum mechanics.

After the 1930s, the logical empiricism of the Vienna Circle became the received view in philosophy of science. Many logical empiricists regarded the debate around realism as metaphysics, and therefore a pseudo-problem without cognitive meaning, so scientific realism was for a long time not a major focus in philosophy of science. Not until the 1960s did a significant number of philosophers of science begin to defend scientific realism again, but since then scientific realism has indeed become one of the central issues in philosophy of science.

Maxwell's challenge to the observational–theoretical dichotomy

Grover Maxwell (1918–1981) was a professor at the University of Minnesota who, in an important paper entitled "The Ontological Status of Theoretical Entities," challenged the traditional observational–theoretical dichotomy (Maxwell 1962, pp. 3–15). In philosophy of science, anti-realists generally hold to the realist position on observational terms, but deny the reality of theoretical terms. So, for anti-realism, the observational–theoretical dichotomy is very important. Maxwell, however, tried to deny the ontological status of the observational–theoretical dichotomy, paving the way for a realist position on all scientific terms, both theoretical and observational.

Maxwell started with an analysis of the concept of "observable." What is observable? Obviously, we can directly see scenery with our naked eyes. But if we see the scenery through the glass of a window, is it still observable? Maxwell further asked, if nearsighted people can only see certain objects with the aid of eyeglasses, are those objects they see observable? How about the satellites that people see through telescopes, the cells that scientists see through microscopes, or the crystal structures seen only with high-energy electron microscopes? Maxwell concluded that there is continuity between the observable and the unobservable, so it is impossible to draw a clear-cut line between observable and unobservable, or between observational terms and theoretical terms.

Maxwell also gave an example of observing cells through microscopes. People used to think that cells, viruses, and other microorganisms could not be observed, but with the invention of the microscope, these microorganisms have become observable. There are two attitudes that might be taken on account of the example: one is that no object is unobservable in principle because, with the progress of science and technology, we may eventually find a way to observe it; the other is that cells are unobservable because seeing through a microscope is not direct observation. Maxwell criticized the latter attitude, which would just as easily make the scenery seen through a window or the objects seen through eyeglasses become unobservable "theoretical terms."

In light of this, Maxwell proposed that (1) there should be no obvious boundary between theoretical terms and observational terms, for the distinction between the two kinds of term depends on the context and on the development level of scientific theories. For example, "cell" used to be a theoretical term, but with the invention of the microscope, it became an observational term. Similarly, microscopic particles at the atomic level used to be unobservable theoretical terms, but with the invention of the electron microscope and other high-energy microscopes, it is possible to observe them. (2) The continuity of theoretical terms and observational terms shows that the distinction between them is arbitrary and has no ontological status.

Thus, Maxwell claimed that what is observable is a theoretical problem itself. The traditional observational–theoretical dichotomy has no ontological status: There is no entity that is unobservable in principle. Since observable things are real, and there is no absolute boundary between observational and theoretical, there is no object that cannot be observed in principle. The distinction between observational and theoretical terms cannot provide an adequate basis for determining whether they are real or not.

Some philosophers have suggested replacing "observable" with "observed." Then, so-called observational terms will refer to objects that have been observed, while theoretical terms will refer to objects that we cannot observe at present. Maxwell argued that such a change would not help, however, because many Inuit people have never seen deserts, and tropical residents may have never seen snow, which makes "deserts" a theoretical term for Inuit people and "snow" a theoretical term for tropical residents. On assessment, such a definition certainly seems unacceptable.

In summary, Maxwell sought to undermine the theoretical basis of antirealism by dissolving the observational–theoretical dichotomy of "observation theory," and his work has had great influence, even if it is not without certain problems.

Van Fraassen's constructive empiricism

Bas C. van Fraassen was born in the Netherlands on April 5, 1941 and immigrated to Canada in 1956. After receiving his bachelor's degree from the University of Alberta in 1963, he moved to the United States, where he studied with Wilfrid Sellars and Adolf Grünbaum at the University of Pittsburgh and received his Ph.D. in 1966. Later, he became a professor at Yale University, Indiana University, the University of Toronto, the University of Southern California, Princeton University, and now San Francisco State University. In 2012, he was awarded the first Hempel Award by the Philosophy of Science Association, recognizing lifetime scholarly achievement. His major works include *An Introduction to the Philosophy of Time and Space* (1970), *The Scientific Image* (1980), *Law and Symmetry* (1989), *Quantum Mechanics: An Empiricist View* (1991) and *The Empirical Stance* (2002). His *Scientific Image*, which received the first Lakatos Award, is a famous book on scientific explanation and scientific realism.

Van Fraassen has given his own precise definition of scientific realism: "Science aims to give us, in its theories, a literally true story of what the world is like; and acceptance of a scientific theory involves the belief that it is true" (van Fraassen 1980, p. 8). Consequently, there are two kinds of anti-realism corresponding to the above definition: one is to believe that scientific theories are true, not in the literal sense, but in the pragmatic sense; the other is to understand truth in the literal sense, so it is unnecessary to think that scientific theories are good because they are true. Van Fraassen chose the latter.

From the pragmatics of explanation, van Fraassen argues that theory choice also has a pragmatic dimension. Accepting a theory involves a commitment to use it when answering questions. It does not really matter whether a commitment is true or not, but whether it can be vindicated. So accepting a scientific theory simply means that we believe it is empirically adequate, even if we know nothing about its truth. He calls this philosophical position "constructive empiricism": Science aims to give us theories that are empirically adequate, and acceptance of a theory involves belief only that it is empirically adequate (van Fraassen 1980, p. 12).

Van Fraassen believes that the main purpose of scientific theories is to "save the phenomena," a notion that was originally put forward by Plato. In ancient times, the observed star trajectories seemed to be very disordered, with the planets sometimes advancing and sometimes retreating, so it was very difficult to obtain a unified explanation. Because of this, Plato sought to find a model into which all the observed phenomena would fit. In this way, the term "save the phenomena" has a kind of instrumental sense. Later, Ptolemy used an epicyclic model – in which every planet revolves uniformly along an epicycle, whose center revolves around the Earth along a deferent – to describe motions of celestial bodies.

From the standpoint of constructive empiricism, van Fraassen provides several arguments against realism.

Theory and observation

For one thing, van Fraassen acknowledges Maxwell's criticism of the observational–theoretical dichotomy, and he agrees that the dichotomy has some degree of ambiguity and arbitrariness. However, van Fraassen believes this merely shows that the "observable" is a vague predicate. Just as we use vague predicates like "old" in our daily lives, so we can safely use the concepts of "observable" and "unobservable."

According to Maxwell, nothing is unobservable in principle. In this regard, van Fraassen proposes that what is observable should be based on the natural limitations of human beings. For example, people can only see light waves of a specific wavelength and hear sounds within a specific frequency range. Other human senses are also limited by our sensory organs. Therefore, "observable" should be strictly understood as "observable to us." Because of the limitations of human organs, some terms are observable to us, but some are not.

For example, what people see through glasses is real, and what we see through telescopes and microscopes can also be considered "observable." But microscopic particles are not observable. Scientists can observe some trajectories in the cloud chamber, but it is not the microscopic particles themselves they are viewing, but the gas trajectories. In the same way, people can observe objects more than hundreds of nanometers in width through optical microscopes, but smaller particles are not observable. This is because, even if we use a high-energy electron microscope, we can only see some diffraction patterns, which cannot count as direct observation. Those unobservable microscopic particles are theoretical terms.

Because van Fraassen retains the observational–theoretical dichotomy, he still takes an anti-realist stance on theoretical terms.

Inference to the best explanation

The rule of inference to the best explanation (IBE) was mainly raised by Gilbert Harman: If a hypothesis H provides the best explanation for some evidence E, then the hypothesis H is probably true. For example, if there are mouse bites on a piece of bread and mouse footprints nearby, then "There is a mouse" provides the best explanation of the evidence, and we can thus infer "There is a mouse."

Van Fraassen rejects the argument from IBE as a justification for scientific realism, however. In his view, on the one hand, people do not necessarily follow IBE; on the other hand, IBE only applies to observable objects, but it cannot justify the existence of microscopic particles. Moreover, IBE has only pragmatic value, but it bears no relation to the truth.

Microstructural explanations and scientific realism

Wilfrid Sellars (1912–1989), one of van Fraassen's teachers at the University of Pittsburgh, held that modern science needs microstructures to explain observable phenomena, so science must believe in unobservable microstructures.

Sellars was born in 1912 in Ann Arbor, Michigan. His father, Roy Wood Sellars, was also a famous philosopher, so Wilfrid Sellars was very interested in philosophy from a young age and inherited his father's critical realism and evolutionary naturalism. Sellars studied mathematics, economics, and philosophy at the University of Michigan in 1931 and then moved to Oxford University in 1934 on a Rhodes scholarship, receiving a bachelor's degree in 1936 and a master's degree in 1940. He began teaching at the University of Iowa in 1938, went to the University of Minnesota in 1947, then to Yale University in 1958, and became a professor of philosophy at the University of Pittsburgh in 1963, where he taught until his retirement in 1982. His main works include *Science, Perception and Reality* (1963), *Philosophical Perspectives* (1967), *Science and Metaphysics* (1968), *Essays in Philosophy and Its History* (1975), *Naturalism and Ontology* (1979), *The Metaphysics of Epistemology* (1989), and *Empiricism and the Philosophy of Mind* (1997). He passed away in Pittsburgh on July 2, 1989.

For our purposes here, Sellars was critical of the traditional positivist view, which divides science into three levels: fact, such as "This raven is black"; empirical laws, such as "All ravens are black"; and theory, such as biological theories. Traditional positivism holds that observable facts can be explained by observable empirical laws, while observable empirical laws can be explained by higher-level theoretical hypotheses. High-level theories are not necessarily restricted to the domain of observables in this view.

For Sellars, however, high-level theories do not explain or contain observable empirical laws. Rather, they only show why observable objects obey such laws to some extent. For example, molecular dynamics cannot explain that the boiling point of water is 100°C; nonetheless, it can show that one of the conductions of the boiling point of water is equal to 100°C if the external pressure is normal atmospheric pressure. As a result, Sellars believed that in modern science, merely resorting to observable empirical descriptions is insufficient, and it is necessary to introduce the unobservable microstructures behind observable phenomena. We need microstructures to account for observations, so science must believe in unobservable microstructures. For example, in order to explain the zero resistance phenomenon of a superconductor, we need to introduce the concept of a "Cooper electron pair." It is difficult (and perhaps inconsistent) for scientists not to believe in the unobservable microscopic structures, yet to believe in the observable phenomena they bring about.

Van Fraassen posits that some scientific phenomena do not need further explanation. For example, according to quantum mechanics, the distribution of microscopic particles has a certain probability. Einstein believed that "God does not play dice with the universe," so he tried to find the explanation behind the probability phenomenon, and later David Bohm put forward the concept of "hidden variables." However, van Fraassen has argued that the Copenhagen interpretation of quantum mechanics needs no further explain the probability phenomenon.

Van Fraassen also suggests that the aim of science is to provide "imaginative pictures" to help us summarize the regularities of observed phenomena. For example, quantum mechanics stipulates that there can only be one particle for each eigenvalue, but in fact there are two electrons in the same energy level. Wolfgang Pauli finally accepted the picture of "electron spin" and imagined that each electron had different spin directions. Although two electrons may have the same energy level, they still correspond to different eigenvalues. Because we are not able to directly observe an electron, we certainly cannot see the spin phenomenon of electrons. Nevertheless, Pauli's concept of "electron spin" has successfully provided an imaginary picture to explain why there are two electrons in the same orbit.

Similarly, John Bardeen, John Schrieffer, and Leon Cooper introduced the concept of a Cooper electron pair, J. J. Thomson developed the "plum pudding" model of the atom, and Ernest Rutherford put forth a planetary model of the atom. All these models are not necessarily true, but they provide imaginary pictures for us to better understand scientific theories.

Moreover, van Fraassen has argued that all introduced microstructures will eventually produce observable results. For example, if different samples of gold dissolve at different rates, even under the same conditions, we may think that the gold samples are mixtures of two substances, A and B, with different melting points. Because the proportions of A and B are different, the melting points of the gold samples are also different. This explains the observable phenomenon of the melting points of gold samples with the unobservable microstructure A and B. In fact, the introduction of microstructures will bring about observable results. We assume that the melting point of A is x, and the melting point of B is x+y. Since the gold samples are mixtures of A and B, we can assert that the melting point of the gold samples should be greater than x but less than x+y. This is an observable phenomenon. Therefore, van Fraassen believes that the introduction of microstructures ultimately explains the observable results, which is for the purpose of empirical adequacy. If the introduction of microstructures does not produce any new experience, it will violate the trend of science, just like the pursuit of hidden variables. But the microstructure in science does not have to be true if it can provide an imaginary picture that is helpful for people to understand. The ultimate aim of science is to find new and observable empirical laws to correct the wrong empirical laws that currently exist.

Limits of the demand for explanation

In his book *Between Science and Philosophy*, J. J. C. Smart draws a distinction between correct theory and useful theory. He asks: How can we accept the usefulness of a theory, but not believe it is true? For example, the captain steered a ship, according to the nautical chart, successfully through the dangerous shoals and reefs and finally arrived at the port safely. How, then, can we imagine that such a successful nautical chart is not true (Smart 1968)?

Van Fraassen has argued against Smart that the validity of science means its theoretical terms are empirically adequate. Thus, science is empirically successful insofar as it can "save the phenomena." Van Fraassen has also used the nautical chart as an example: Sometimes the locations of dangerous shoals and reefs would be exaggerated for the sake of safety. Such a chart is not strictly true, but it is safe for us to drive according to it, so it is still a successful map.

Smart also believed that we must have correct theories to explain the success of theories. Imagine a person in the sixteenth century who was a realist with respect to Copernicus' heliocentrism and instrumentalist with respect to the Ptolemaic system. He would have been able to demonstrate the instrumental validity of the Ptolemaic system because it could produce predictions of planetary motion that were fairly similar to Copernican theory. Therefore, the reality of Copernican theory justifies the validity of the Ptolemaic system. If all theories are regarded as purely instrumental, it is impossible to explain the instrumental validity of a particular theory. Van Fraassen's response is that fundamental regularities are always brute regularities, which cannot be further explained. Therefore, it is unnecessary to provide a realistic explanation for the instrumental theory.

Smart also raised the problem of cosmic coincidence: The observable regularities must be explained by the theoretical terms of a deeper level; otherwise, are the observed regularities in the world only a kind of accidental coincidence? Van Fraassen's response to this is that cosmic coincidence does not mean there is no explanation. For instance, if I meet someone in the market by chance, I can explain why I am going to the market at this time, and the person can also explain why he/she goes to the market at this time. These reasons add up to explain why we have met in the market. We call it "coincidence," then, not because the event cannot be explained, but because I did not meet the person on purpose. Therefore, the regularities displayed by the universe can be explained respectively by the scientific laws, not necessarily by a unified explanation at a deep level.

In summary, van Fraassen has concluded that there are limits to the demand for scientific explanations. He objected to the unlimited demand for scientific explanations by philosophers such as Smart, Reichenbach, Salmon, and Sellars. Because the unlimited demand for scientific explanations may lead to the hidden variable explanation of the probability phenomena at the quantum level – which violates the Copenhagen interpretation, the dominant view in quantum physics in the twentieth century – van Fraassen insists on an empirical stance, while rejecting scientific realism.

Darwinian explanation of the Ultimate Argument

To defend realism, Hilary Putnam proposed the Ultimate Argument, often called the "no miracles argument." It runs as follows: If we do not believe that our theories are true and that our theoretical terms refer to unobservable objects, then we must admit that the success of our science is a miracle (Putnam 1975, p. 69). After all, it is exceedingly difficult for us to tell a lie that can be successful from beginning to end. For example, in the famous fictional tale *The Siege of Berlin*, Alphonse Daudet describes how a kind girl, conspiring with a doctor, tried to persuade her grandfather, a veteran French colonel, that France had defeated Prussia in the Franco–Prussian War. The girl and the doctor arranged the lie very delicately and achieved a certain amount of success – so much so that the old colonel believed France had really defeated Prussia, and his health gradually improved. In the end, however, the Prussian army entered Paris, and the colonel died of despair after seeing the scene from his balcony. Likewise, there are so many phenomena in the world around us that we can readily observe that if modern science were not real, but could still successfully explain all those observed phenomena, it really would be a miracle. In this way, Putnam uses his "no miracles argument" to defend realism.

In response, van Fraassen provides a Darwinian account of the success of modern science. When we look at nature, we often marvel at the delicacy of animals and plants: the rapid speed of cheetahs, the long necks of giraffes, the keen smell of dogs, and the beautiful blooming of roses. From this, some people may think that the subtlety of nature is either a miracle or God's meticulous design. Yet according to Darwin's theory of evolution, all animals and plants are actually the result of competition for survival and natural selection. If cheetahs did not have such speed, they would not be able to hunt for food; if giraffes did not have such long necks, they could not eat the leaves on the treetops; if dogs did not have such a keen sense of smell, they would not be able to find prey or avoid certain dangers; if roses did not have such beautiful flowers, they could not provoke insects to help pollinate them. In this way, the delicacy of nature need not be seen as a miracle or God's design, but as the result of natural selection.

Analogously, van Fraassen thinks the same is true of the success of modern science. In fact, there have been countless theories attempting to explain observable phenomena in the history of science, but only successful theories have survived. The reason why the theories of modern science are so successful is that the failed theories have been weeded out and only the successful theories have survived. Therefore, the success of science is not a miracle, but the result of natural selection as well. Indeed, as van Fraassen sees it, the "no miracles argument" is helpless to defend scientific realism.

Laudan's confutation of convergent realism

There is also a kind of realism called "convergent realism," which holds that the cognitive process of human beings is constantly approaching the truth. For example, Popper put forward the concept of verisimilitude under the pretense that the development of science is the process of approaching truth through the method of hypothesis falsification. Larry Laudan has also regarded Richard Boyd, W. H. Newton-Smith, Hilary Putnam, Michael Friedman, and Ilkka Niiniluoto as convergent realists.

Because convergent realism tends to focus on the epistemological perspective, Laudan called it "Convergent Epistemological Realism" (CER) and summarized the claims of CER in this way (Laudan 1981, pp. 21–22):

R1 Scientific theories are typically approximately true, and more recent theories are closer to the truth than older theories in the same domain. For example, the theory of relativity is closer to the truth than Newtonian mechanics in the domain of mechanical phenomena.

R2 Observational and theoretical terms within the theories in a mature science genuinely refer – i.e., there are substances in the world that correspond to the ontologies presumed by our best theories. For example, the subatomic particles postulated in modern science really exist.

R3 Successive theories in mature science will preserve the theoretical relations and apparent referents of earlier theories – i.e., the latter will be the limiting cases of the former. For example, the theory of relativity preserves many concepts in Newtonian mechanics; furthermore, the latter can be seen as a limiting case of the former when the speed is much less than the speed of light, $v<<c$.

R4 Accepted new theories do and should explain why their predecessors were successful insofar as they were successful. For example, the theory of relativity can explain the success of Maxwell's equations of electrodynamics by resorting to the relativistic effects of charge movements.

R5 The above theses entail that mature scientific theories should be successful; indeed, the theses constitute the best explanation for the success of science. The empirical success of science accordingly provides striking empirical confirmation for realism.

As an empiricist, Laudan criticized convergent realism, believing that cases in the history of science show that there is no necessary link between the success and the verisimilitude of science. He first analyzed the "downward path" of scientific theories – that is, whether an approximately true theory whose theoretical term really refers must be a successful scientific theory.

Laudan criticized convergent realists for never providing explicit definitions of the concepts "verisimilitude," "approximately true," or "close to truth," although Popper did propose an understanding of verisimilitude as the true content of a theory being larger than the false content, such that (Popper 1963, p. 220):

$$CtT(T_1)>>CtF(T_1)$$

But if from Popper's definition we cannot logically deduce that the content of the theory is approximately true, then it is probable that all those untested parts are wrong.

Even if the convergent realists can explicitly define the concept of verisimilitude, from the perspective of the history of science, many theories with real references are not actually successful. For example, the concept of the atom in ancient atomism does refer, but ancient atomism was for a long time unsuccessful, even to the point that it disappeared in Europe for a time. Not until John Dalton put forward modern atomic theory was atomism revived. Thus, historically speaking, even an approximately true theory is not necessarily successful.

Furthermore, we can construct any theory with referential concepts, such as electrons and atoms, in principle. Since these theories are arbitrarily constructed, we cannot expect them to be successful. Therefore, theories with true references are not equal to successful theories.

From the upward path, are successful theories approximately true in the history of science? Laudan also cited a large number of counterexamples, such as the crystalline spheres model of ancient and medieval astronomy; the humoral theory of medicine; the effluvial theory of static electricity; catastrophist geography; the phlogiston theory of chemistry; the caloric theory of heat; the vibratory theory of heat; the vital force theories in physiology; the theory of electromagnetic ether; the theory of optical ether; the theory of circular inertia; and theories of spontaneous generation (Laudan 1981, pp. 43–45). Taking the ether theory as an example, the wave theory of light introduces the concept of optical ether to explain the propagation of light waves in a vacuum. In this theory, it was believed that the ether is omnipresent in the universe and is the medium of light waves. The concept of ether made the wave theory of light a great success in the eighteenth and nineteenth centuries, and it was later introduced into electromagnetics and used even more widely. In fact, it was not until the theory of relativity that physicists gave up the concept. As this demonstrates, the ether is not real, but the theory of ether achieved great success, leading Laudan to believe that successful scientific theories are not necessarily approximate to the truth, and their theoretical concepts genuinely refer.

In summary, both the downward path and the upward path show that the connection between the verisimilitude and the success of a scientific theory is much weaker than convergent realists may believe. Laudan thus draws eight conclusions:

> (1) The fact that a theory's central terms refer does not entail that it will be successful; and a theory's success is no warrant for the claim that all or most of its central terms refer. (2) The notion of approximate truth is presently too vague to permit one to judge whether a theory consisting entirely of approximately true laws would be empirically successful; what is clear is a theory may be empirically successful even if it is not approximately true. (3) Realists have no explanation whatever for the fact that many theories which are not approximately true and whose theoretical terms seemingly do not refer are nonetheless successful. (4) The convergentist's assertion that scientists in a mature discipline preserve, or seek to preserve, the laws and mechanisms of earlier theories in later ones is probably false;

his assertion that when such laws are preserved in a successful successor, we can explain the success of the latter by virtue of the truthlikeness of the preserved laws and mechanisms, suffices from all the defects noted above confronting approximate truth. (5) Even if it could be shown that referring theories and approximately true theories would be successful, the realists' argument that successful theories are approximately true and genuinely referential takes for granted precisely what the non-realist denies (namely, that explanatory success betokens truth). (6) It is not clear that acceptable theories either do or should explain why their predecessors succeeded or failed. If a theory is better supported than its rivals and predecessors, then it is not epistemically decisive whether it explains why its rivals worked. (7) If a theory has once been falsified, it is unreasonable to expect that a successor should retain either all of its content or its confirmed consequences or its theoretical mechanisms. (8) Nowhere has the realist established—except by fiat—that non-realist epistemologists lack the resources to explain the success of science.

(Laudan 1981, pp. 21–22)

Hacking's experimental realism

Ian Hacking received his bachelor's degree from the University of British Columbia in 1956 and his Ph.D. from Cambridge University in 1962. He has been a professor in the Institute of History and Philosophy of Science and Technology at the University of Toronto and a fellow of the Royal Society of Canada, the British Academy, and the American Academy of Arts and Sciences. He won the Killam Prize for the Humanities, the highest academic honor in Canada, in 2002, and the Order of Canada in 2004. According to Hacking, experimental sciences provide the strongest evidence for scientific realism (Hacking 1982, pp. 71–87).

Concerning the debate of scientific realism, Hacking thinks that the problem cannot be solved at the theoretical level; nevertheless, he finds that in experimental activities, scientific realism is inevitable, so he proposes experimental realism. Hacking further distinguishes two kinds of realism: realism about theories and realism about entities. Theory realism affirms that scientific theories are real descriptions of the world, while entity realism claims that the main concepts in natural science, such as waves, fields, and black holes, really exist.

With that said, theory realism and entity realism hold something in common. Generally speaking, accepting a theory as true means that the main concept of the theory is also referential, but theory realism and entity realism are not identical. As Hacking points out, many experimental physicists are entity realists who believe that the theoretical entities they study are real, but they are not theory realists because they tend to take an instrumentalist position with respect to theoretical models.

The experimental realism put forward by Hacking is essentially entity realism: Most experimental physicists believe in the existence of unobservable entities, but they may doubt whether their theories are true. Here, Hacking borrows Putnam's referential model of meaning: Meaning is a vector, just like a dictionary, in which first comes the syntactic marker (such as whether the word is a noun or an adjective), next the semantic marker (i.e., the general category of the word), then the stereotype (i.e., the standard usages of the word), and finally the reference. When the category changes, the stereotypes of the word may be modified, but the reference can remain the same. So even the stereotype of a concept may change and its meaning may be different, but it can still refer to the same thing.

Taking the concept of the electron in the history of science as an example, J. J. Thomson once imagined that electrons were evenly distributed in atoms; Bohr's model of the electron was similar to a planetary orbit; Robert Millikan accurately measured the value of electric charge; and, furthermore, our current understanding of electrons entails an electronic cloud subject to probability distribution. Thus, throughout the history of science, the stereotype of the concept of the electron has changed constantly, and its meaning has also changed, so does the concept refer to the same thing?

In Hacking's view, the concept of the electron was first introduced to explain some electrical phenomena. Later, the understandings of Thomson, Bohr, and Millikan on the concept were based on certain scientific laws and explained some specific effects. Although the stereotype of the electron is constantly changing, the reference of "electron" remains the same, as it still explains the original electrical phenomena.

Hacking uses experimental arguments to defend realism, maintaining that whether a theoretical concept really exists depends on whether we can use it to study other theoretical concepts or complex phenomena. For example, the reality of the concept of the electron does not lie in the fact that we can successfully deduce "electron" from some experimental results, nor that we can accurately predict some experimental effects only by believing in the electron; rather, when we start to build – and often successfully build – new devices on a regular basis, these devices make good use of the properties caused by the electron to intervene in the uncertain nature, such that we are completely convinced of the reality of the electron.

Hacking takes PEGGY II as an example. The design of this project uses the concept of the electron, but very little of the theory of electrons. Even if modern quantum electrodynamics were completely modified, PEGGY II could still work. In 1978, *The New York Times* reported experimental results of PEGGY II: (1) Parity is not conserved in the scattering of polarized electrons from deuterium; (2) parity is violated in a weak neutral current interaction. Since we have successfully used the concept of the electron to study the phenomena of parity conservation, the electron really exists.

Thus, the reality of a theoretical term, to a certain extent at least, depends on the development of human knowledge. We now think that electrons are real because we can use electrons to study other microscopic particles. But what about whether neutral bosons really exist? Hacking thinks this is not yet determinable. When we can use neutral bosons to study other things, neutral bosons will no longer be a hypothetical concept, but as real as an electron.

Therefore, the ontological status of theoretical entities has a temporal dimension. Hacking thinks that scientists in the nineteenth century could regard the atom as a fiction because they mainly used the concept of atoms to explain the chemical reactions of substances. But since the twentieth century, because scientists can now use the concept of the atom to design delicate experiments to study more microscopic particles, scientists have taken a realist position on the concept of the atom.

Regarding the experimental argument for entity realism, Hacking writes:

> The direct proof of electrons and the like is our ability to manipulate them using well understood low-level causal properties...We can, however, call something real...only when we understand quite well what its causal properties are...Hence, engineering, not theorizing, is the proof of scientific realism about entities.
>
> (Hacking 1982, pp. 71–87)

Fine's "Natural Ontological Attitude"

Arthur Fine was born in Lowell, Massachusetts on November 11, 1937. He received his bachelor's degree in 1958 and his Ph.D. in 1963 from the University of Chicago. He has taught at Cornell University, the University of Illinois at Chicago, and Northwestern University, and is now an emeritus professor at the University of Washington. His monographs include *The Shaky Game: Einstein, Realism and Quantum Theory* (1986) and *Bohmian Mechanics and Quantum Theory* (1996).

Fine has criticized realism from three aspects: (1) the overall argumentation strategy; (2) the problem of a small handful; and (3) the problem of conjunction in approximate truth (Fine 1984, pp. 83–107). He also points out that, according to David Hilbert's proof theory, metatheoretical arguments must meet more stringent requirements than the issue to be discussed. Therefore, if we want to argue for realism, we must adopt a more rigorous method than scientific practice.

Fine criticizes the overall argument for realism of merely begging the question. Because realism claims that, at the ground level, theoretical models in science can produce new confirmed predictions and unify seemingly separated phenomena or fields, only realism can explain the success of science; at the level of methodology, only realism can explain how science can succeed under the guidance of methodology, such as Popper's method of hypothesis falsification. But the argument needs to add a premise: We must regard a good explanatory hypothesis as true. The premise of realism is also the conclusion of the argument, so it commits the fallacy of begging the question.

In addition, realism has the problem of a small handful. Realism holds that in any field of science, the number of the candidate theories in a certain period of time is always quite limited. Moreover, among the small handful of theories, there is a family resemblance, and the new theory often keeps the content of the old one. The "small handful" strategy is very successful in science. Realism claims that this illustrates that the scientific theory is approximately true, so that the number of candidate theories is small and the new theory has family resemblances with the old one.

Fine raises three questions about the problem of a small handful: (1) Why is there only a small handful out of the theoretically infinite number of possibilities? (2) Why is there the conservative family resemblance between members of the handful? (3) Why does the strategy of narrowing the choice in this way work so well? Furthermore, Fine believes that realism has not answered the first question, that the answer to the second question involves circular argumentation, and that the answer to the third question lacks evidential support. In fact, we can say to the contrary: (1) Theory choice in science has many possibilities, not a small number; (2) there will also be sharp ontological changes in science, such as how the theory of relativity abandoned the concept of ether without necessarily having any family resemblance between the new theory and the old one; and (3) the strategy of a small handful often fails, while only occasionally succeeding. But does realism fail to explain why the strategy that usually fails can occasionally succeed?

Fine also raises the problem of conjunction, namely, if T and T' are mutually independent and well-confirmed theories with explanatory force, and there is no ambiguity between their common terms, then the conjunction $T \wedge T'$ has the explanatory force of the two theories, and it should also be a reliable instrument for prediction.

But according to the realistic notion of approximate truth, "T is approximately true" and "T' is approximately true," they will both have a certain deviation. If the deviation of theory T and that of T' is ε, then the deviation of $T \wedge T'$ should be 2ε, such that the deviation of $T \wedge T'$ is larger than that of theory T and theory T' alone. According to the criterion of approximate truth, $T \wedge T'$ should be more unreliable than a single theory and therefore must be abandoned. This contradicts the previous conclusion and is thus unacceptable to scientists in practice.

Arguing against realism, however, Fine is unwilling to accept anti-realism, so he tries to chart a third way, adopting attitudes of minimalism, common sense, and non-realism toward science. This he calls the Natural Ontological Attitude (NOA).

Fine thinks both realism and anti-realism uphold a core position: accepting scientific truths. What distinguishes realism from anti-realism is what they add to the core position. Realism adds the external direction to NOA – that is, the correspondence between the external world and the approximate truth; anti-realism adds the internal direction – that is, how to reduce scientific truth, scientific concepts, and scientific explanation to our experience.

In philosophy, Fine accepts minimalism or deflationism, as minimalism can result from deflationism taken to its maximum degree (Horwich 1990). He claims that "perhaps the greatest virtue of NOA is to call attention to just how minimal an adequate philosophy of science can be" (Fine 1984, pp. 83–107). He therefore claims his NOA takes the minimum common divisor of realism and anti-realism to get the core position.

Strictly speaking, NOA is neither realism nor anti-realism, but rather moderate non-realism. When NOAers accept scientific truth, they believe that the entities, properties, relations, and processes postulated by scientific theories are also true. However, NOAers do not commit to the progressivism of realism, remaining open to paradigm shifts or concept revolutions in science.

NOA also emphasizes that philosophers of science should respect the scientific research of practicing scientists and believe in the achievements of scientific research, without providing metaphysical restrictions for science or offering unnecessary explanations. Therefore, NOA seeks to find a middle way between realism and anti-realism.

Musgrave's defense of realism

Alan Musgrave received his Ph.D. from the London School of Economics, studied with Sir Karl Popper, and has been teaching at the University of Otago since 1970. His writings include *Common Sense, Science, and Scepticism* (1994) and *Essays on Realism and Rationalism* (1999). The collection *Criticism and the Growth of Knowledge*, coedited by Musgrave and Lakatos, has also become a classic in the field of philosophy of science.

Musgrave systematically argues against both van Fraassen's constructive empiricism and Fine's NOA. He compares the risks, losses, and benefits of realism and constructive empiricism, finding that the empirical adequacy required by constructive empiricism for scientific theory is not only appropriate now, but also must be appropriate in the future. Scientific theory has to save all phenomena, so constructive empiricism takes the same risks as realism. Since constructive empiricism and realism bear the same risks, and realism has more benefits, with commitment to the truth of theory, the rational choice should be realism.

Musgrave also criticizes some of van Fraassen's arguments for anti-realism. First of all, constructive empiricism, he contends, cannot explain the observational–theoretical dichotomy. Because, for constructive empiricists, the statement "B is unobservable to humans" contains the unobservable term B, it is impossible to determine whether it is true. Second, constructive empiricists' limitation of the demand for scientific explanation actually confuses realism and essentialism, as well as "demand for explanation" and "demand for ultimate explanation." Only in the Aristotelian view of essentialism must a genuine explanation be final or self-explanatory. It is obviously wrong that van Fraassen has deduced that science cannot explain anything from his doubt of ultimate explanation. For example, the Schrödinger equation in quantum mechanics is a fundamental law about which we cannot get a further explanation, but we can use the equation to explain most effects in quantum physics. Third, van Fraassen's Darwinian explanation of scientific theories can only explain why scientific theories are successful in general, but not why a particular scientific theory is successful. In fact, we can ask two questions: (1) Why can a particular theory succeed? (2) Why can scientific theories succeed in general? Van Fraassen can only explain the latter, but does not answer the former, so the Ultimate Argument, or "no miracles argument," may in fact still hold.

Musgrave points out that most scientists still swear loyalty to realism, being eager to understand the world and pursue the illustrative truth. Realism regards theoretical science as trying to understand the world, and there is continuity between common sense and scientific knowledge, whereas constructive empiricism seeks to drive a wedge between theoretical science and common sense. Of course, Musgrave also admits that van Fraassen's constructive empiricism is more convincing than earlier anti-realist arguments, which often left him sleepless; however, Musgrave believes that constructive empiricism is weaker than earlier anti-realism and thus correspondingly closer to realism. His hope is thus "that realism emerges a little bloodied but unbowed from its encounter with constructive empiricism" (Musgrave 1985, pp. 197–221).

Musgrave also disagrees with Fine's NOA. Different schools of thought may have different understandings of the concept of truth – by which, for example, the positivists mean "useful," the constructive empiricists mean "empirical adequacy," and the realists mean "correspondence to reality." Since different schools of thought have different understandings of truth, they accept scientific truth in different senses, so they may have different "core positions." NOA expects different philosophical schools to share a common "core position," but Musgrave argues that there is actually no such "core position."

Fine's NOA advocates for philosophy of science to accept the opinions of scientists and for philosophers of science to accept what scientists believe. Because the two fundamental theories of modern science – the theory of relativity and quantum mechanics – are both inclined to positivism and instrumentalism, Fine's philosophical position is close to anti-realism. However, Musgrave charges that Fine's NOA actually uses "Davidsonian-Tarskian referential semantics," which already entails a thoroughly realist position. As a result, Musgrave has humorously quipped, "NOA's Ark will after all be Fine for realism" (Musgrave 1989, pp. 383–398).

Summary

Scientific realism has been a very hot topic in philosophy of science in recent years. In 2000, Stathis Psillos at the University of Athens wrote "The Present State of the Scientific Realism Debate," which introduces the latest developments of scientific realism, showing just how vibrant the debate is. Indeed, there are many philosophers in the respective camps of scientific realism and anti-realism who have put forward sophisticated arguments to defend themselves. Here, however, the author tries to look at the debate from another perspective.

The central debate between scientific realism and anti-realism focuses on whether the theoretical terms in science are real or merely human constructs. Therefore, arguments about scientific realism are important because they may provide us with new arguments and better ideas for understanding scientific concepts. In a reference entry on "Scientific Realism and Antirealism," Fine points out the understanding of science from different philosophical standpoints: Realists hold that scientific truth is independent of the observers; internal realists hold that scientific truth is relative to our conceptual schema; constructive empiricists accept scientific theories only because they are empirically adequate; and NOAers only accept scientific theories (Fine 2000). Thus, all the debates about realism also contribute to our understanding of science itself.

Generally speaking, most of the anti-realists in the debate over scientific realism have inherited the tradition of logical empiricism, so they are opposed to the introduction of metaphysics into science. In this discursive context, realists defend themselves but are unwilling to make metaphysical commitments. If there is to be a novel approach in the debate moving forward, it may be necessary to break through this empirical framework. The author tends to hold an open attitude as to whether physical science should come into greater contact with metaphysics.

10 Philosophy of scientific experimentation

So far, the previous chapters of this book have mainly been about scientific theory, with few references to scientific experiments. For example, the meaning criterion is used to discuss whether a theory has cognitive significance; the problem of induction relates to the study of whether or how to derive scientific theory by means of induction; the nature of scientific laws pertains to whether laws of nature are regular or necessary; the model of scientific growth speaks to how scientific theories develop; the purpose of demarcation is to distinguish whether a theory is scientific or pseudoscientific; and the debate of scientific realism focuses on whether a scientific theory is approximately true or not, and whether a theoretical term has a reference. In these chapters, it therefore seems that experiments have rarely been mentioned. If anything, they are often treated in a subordinate position: We generalize theories from experiments, and we argue about whether crucial experiments can provide the final basis for competing theories.

But in real science, scientific experiments play an important role in scientific research. Indeed, the vast majority of scientific research focuses on scientific experiments rather than purely theoretical inquiry (not to mention that finding work in experimental physics is much easier than in theoretical physics). Recognizing the importance of experimentation in science as a whole, this chapter will make a philosophical analysis of scientific experimentation, focusing on the relationship between scientific theories and scientific experiments to answer the big question: Which comes first, theory or experiment?

The view of experimentation in the traditional philosophy of science

Logical empiricists' view of experimentation

The traditional philosophy of science pays relatively less attention to scientific experimentation. In the first half of the twentieth century, logical empiricism was the received philosophical position, as both the Vienna Circle and the Berlin Circle emphasized "logic" and "experience." In this sense, "logic" refers to modern mathematical logic, whereas "experience" is usually expressed by the propositions we can observe directly, such as "This flower is red," "The grass is green," "The scale of that thermometer is 25 degrees," etc. According to logical empiricists, what we observe is certain without doubt. In English, "I see" often means "I understand," and even in Chinese, there is also a saying that "It is better to see once than hear a hundred times." In a similar way, logical empiricists usually prefer "observation" to "experimentation."

Especially according to logical empiricists' syntactic view, scientific theories can usually be reorganized into an axiomatic deductive system (Liu Chuang 1997, pp. 147–164). For example, John von Neumann (1903–1957) outlined the axiomatic system of quantum mechanics in *Mathematical Foundations of Quantum Mechanics* (von Neumann 1955). But how can we know which deductive systems are true? We usually examine whether the observable results derived from these deductive systems are consistent with empirical observations. For example, the axiomatic system of quantum mechanics is very complex, but the semiconductor and superconductivity effects derived from it can be observed directly, so these effects justify the validity of quantum mechanics.

The syntactic view of scientific theories can be drawn with the following picture:

In such a schema, philosophy = logic + experience; science = theory + observation; and the status of scientific experimentation is often ignored. Scientific experiments are mostly used to create or purify our observations, so their status is relatively secondary.

Accordingly, logical empiricists' view of experimentation is relatively simple: We first have observational reports or experimental results, and then we generalize the theoretical knowledge. The development of science is a process in which scientific theories are discovered and accumulated. The disadvantages of such a view of experimentation are that: (1) It will run into the challenge of the problem of induction; (2) it pays much more attention to observations than to experiments; and (3) whether they are observations or experiments, all are considered ultimately for the sake of scientific theories. Still, logical empiricists' view of experimentation respects the independence and objectivity, or intersubjectivity, of observations and experiments, such that when there is any dispute between scientific theories, we can use observations or experiments to make a rational decision.

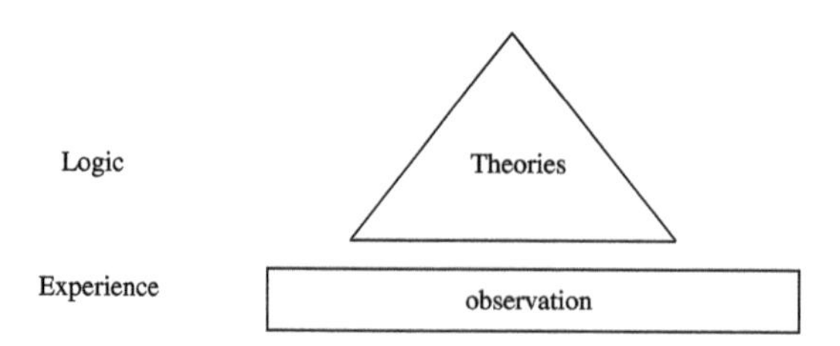

Figure 10.1 Logical empiricists' picture of theory and observation

Popper's view of experimentation

In contrast, the independence and objectivity of observations and experiments were challenged by N. R. Hanson (1924–1967). Hanson studied at Oxford University and Cambridge University, and after graduation he taught at Indiana University, Princeton University, and Yale University. Among other things, he proposed that observations are "theory-loaded" in his masterpiece *Patterns of Discovery* (Hanson 1958). Almost at the same time, moreover, Feyerabend also put forward the claim that observations are "theory-laden."

As a critical rationalist, Popper is also famous for having upheld the thesis of the theory-ladenness of observation. He once asked an audience to observe freely in his lecture, but the audience was confused, illustrating that we must have problems or theories before we can really engage in observation. Later, Alan F. Chalmers, a British philosopher of science, gave a more clear exposition of a similar position in *What Is This Thing Called Science?* (Chalmers 1999). General points in this thesis are as follows:

(1) Visual experience does not completely depend on the images on the retina. For example, in daily life we can see that the four suits of playing cards are red hearts, red diamonds, black spades, and black clubs. But in a card experiment, experimenters may deliberately show some unusual colors, such as black hearts or red spades. If the presentation time is very short, the observer will often mistakenly regard these as being of normal suits, such as black spades or red hearts. Another example is that when new students in a medical school begin to study the X-ray diagnosis of lung disease, they only see some inexplicable light spots and shadows on a fluorescent screen, but after becoming experts through continuous in-depth learning, they come to see the physiological variations and pathological changes in different cases.
(2) Observation must presuppose theories. For example, when we say, "The electron beam is repelled by the north pole of the magnet," we have presupposed theoretical terms such as "electron," "north pole," and "magnet." Another example is that when a teacher says, "This is a piece of chalk," but students suspect it is fake, even if the teacher uses the chalk to draw a white line on the blackboard, the students can still argue that other items can also draw a white line; if the teacher uses a chemical reaction to confirm that its chemical composition is indeed calcium carbonate, then they are still resorting to scientific theories to support the observation.
(3) Theories guide observations and experiments. For example, without the prediction of James Maxwell's electromagnetic wave equations, it is difficult to imagine how Heinrich Hertz could conduct his radio wave detection experiments.

Since observation is always theory-laden, Popper criticized logical empiricists' use of observations or experiments to confirm theories as committing the fallacy of circular argumentation: If observations or experiments themselves have already presupposed theories, then would employing observations or experiments to support those theories not ultimately be using theories to support theories? We can abbreviate this with the following schema:

Figure 10.2 Popper's criticism of logical empiricists' view on theory and experiment

To the contrary, if we use the method of hypothesis falsification, we can effectively avoid circular argumentation: If an observation or experiment itself has presupposed a theory, but the current observation or experiment still does not support the theory, the theory itself is certainly problematic.

Although Popper was against the independence of scientific experimentation, he still affirmed the role of experiments in theory choice – especially that of crucial experiments in directly falsifying scientific theories. For example, we now have competing theories T_1 and T_2. According to Popper, theory T_1 is superseded by T_2 in the sense that T_2 corresponds better to the facts than T_1, if and only if (Popper 1963, p. 232):

T_2 makes more precise assertions than T_1, and these more precise assertions stand up to more precise tests.
T_2 takes account of, and explains, more facts than T_1 (which will include, for example, the above case that, other things being equal, T_2's assertions are more precise).
T_2 describes, or explains, the facts in more detail than T_1.
T_2 has passed tests that T_1 has failed to pass.
T_2 has suggested new experimental tests not considered before T_2 was designed (and not suggested by T_1, and perhaps not even applicable to T_1), and T_2 has passed these tests.
T_2 has unified or connected various hitherto unrelated problems.

A crucial experiment can be used to support T_2 and disconfirm T_1. For example, in the seventeenth century, when it came to theories concerning the nature of light, there were Isaac Newton's particle theory and Christiaan Huygens' wave theory, respectively. Due to Newton's academic contribution and lofty position, the particle theory prevailed in the seventeenth and eighteenth centuries. But in the early nineteenth century, physicists such as Augustin-Jean Fresnel (1788–1827) studied the diffraction of light and revived the wave theory. Interestingly, French mathematician and physicist Siméon Denis Poisson (1781–1840) had pointed out that, if the wave theory was right, then there must be a bright spot in the center of the shadow of a round disk, yet Poisson himself thought this was impossible. Fresnel, however, in collaboration with experimental physicist François Arago, experimentally demonstrated that there is a bright spot in the center of the shadow of a disk – now, quite fittingly, known as "Poisson's bright spot." Since this experiment, the wave theory of light has once again occupied a dominant position, and the particle theory has been falsified.

Thus, in *The Logic of Scientific Discovery*, Popper clearly claims that theories are prior to experiments:

> The theoretician puts certain definite questions to the experimenter, and the latter by his experiments tries to elicit a decisive answer to these questions, and to no others. All other questions he tries hard to exclude…It is a mistake to suppose that the experimenter [aims] "to lighten the task of the theoretician", or…to furnish the theoretician with a basis for inductive generalizations. On the contrary the theoretician must long before have done his work, or at least the most important part of his work: he must have formulated his questions as sharply as possible. Thus it is he who shows the experimenter the way. But even the experimenter is not in the main engaged in making exact observations; his work is largely of a theoretical kind. Theory dominates the experimental work from its initial planning up to the finishing touches in the laboratory.
>
> (Popper 1959, p. 107)

Kuhn, Lakatos, and Feyerabend

With the rise of Kuhnian concepts of paradigm and incommensurability, and the recognition of different scientific standards for different paradigms, there would no longer be crucial experiments for discerning between different paradigms. Later, Feyerabend further developed the concept of incommensurability: For different theories in different paradigms, the meaning of the concept is different, and there is no crucial experiment at all. In his view, since experiments provide no objective test, scientific theories operate only according to the anti-credo of "Anything goes."

Lakatos sought to improve upon Popper's falsificationism, claiming that a scientific theory is not falsified by any crucial experiment, but is replaced by a new theory with more empirical content. Thus, he proposed that a crucial experiment is never decisive; rather, T is replaced by another theory T' if and only if:

> (1) T' has excess empirical content over T, that is, it predicts novel facts – or facts improbable in the light of, if not forbidden by – T; (2) T' explains the previous success of T, that is, all the unrefuted content of T is included (within the limits of observational error) in the content of T'; and (3) some of the excess content of T' is corroborated.
>
> (Lakatos 1978, p. 32)

For example, in the second half of the nineteenth century, astronomers noticed that the orbit of Mercury at perihelion did not match the prediction of Newtonian mechanics. At that time, physicists took the discovery of Neptune and Pluto as examples, conjecturing that there might be an asteroid near Mercury, which was once called "Vulcan." But while astronomers never saw Vulcan in the first place, not being able to observe Vulcan did not directly falsify Newtonian mechanics. At that time, physicists further conjectured that a cloud of cosmic dust nearby had hindered our observation of the planet. It was not until Einstein's general theory of relativity that physicists gave up the conjecture of Vulcan altogether. Thus, it was neither observation nor experiment with respect to the precession data of Mercury at perihelion that falsified Newtonian mechanics; instead, it was because we accepted a theory – the general theory of relativity – that we abandoned the old theory of Newtonian mechanics.

Therefore, ever since the rise of historicism, the independence, objectivity, and even importance of scientific experimentation have been questioned time and time again, as the mainstream within the traditional philosophy of science prefers theories to experiments, insisting that theories come prior to experiments in both logic and importance.

New experimentalism

It was Ian Hacking, the representative of new experimentalism, who first turned to an increased emphasis on scientific experimentation in philosophy of science. We have briefly introduced Hacking and his experimental realism in the previous chapter on scientific realism, but Hacking is in fact a very versatile philosopher who has also made great achievements in the fields of philosophy of scientific experimentation, probability theory, and philosophy of language. His main works include *The Logic of Statistical Inference* (1965), *The Emergence of Probability* (1975), *Why Does Language Matter to Philosophy?* (1975), *Representing and Intervening* (1983), *The Taming of Chance* (1990), *Scientific Revolutions* (1990), *Rewriting the Soul: Multiple Personality and the Sciences of Memory* (1995), *Mad Travelers: Reflections on the Reality of Transient Mental Illness* (1998), *The Social Construction of What* (1999), *An Introduction to Probability and Inductive Logic* (2001), and *Historical Ontology* (2002). Let us explore some of his thoughts on the relationship between theory and experimentation in philosophy of science.

Theory-oriented or experiment-oriented

First of all, Hacking opposes a "theory-oriented" approach and instead advocates for the restoration of experimentation in philosophy of science. In his book *Representing and Intervening*, he first ironically mentions the comparative analogy of the Academy at Athens and the temple of metallurgists. As most of us know, Plato and Aristotle frequented the Academy at Athens, which symbolizes pure theory in Hacking's analogy. However, very few people have heard of the temple to the god of fire, the patron of metallurgists, which symbolizes such impure affairs as experiments. Of all the buildings that once graced the Athenian Agora, however, only the temple of metallurgists has remained standing, untouched by time or reconstruction. The rebuilding of the Academy, which represents pure theory, was funded partly by money from the steel companies in Pittsburgh.

Hacking also compares two physicists, Robert Boyle (1627–1691) and Robert Hooke (1635–1703). Boyle was a theoretician who also experimented. He not only enjoyed a high position in the scientific community, but he was also a noble. By contrast, although Hooke is famous for Hooke's law, he mainly did experimental work. Even though Hooke was the first to build a new reflecting telescope, discovered major new stars, conducted microscopic research of the highest rank, and laid important groundwork for the concept of the cell, comparatively speaking, Hooke's fame in the history of science is much less than Boyle's. Not only that, but Hooke himself was poor and self-taught. Hacking thus uses these two Enlightenment-era contemporaries as an example to make an ironic point: In science, theoreticians are noble, but experimenters are poor.

Two more recent examples of the same phenomenon are the brothers Fritz London (1900–1953) and Heinz London (1907–1970). They were born in Germany and both did physics research. Fritz made significant theoretical contributions to quantum chemistry and superconducting physics, and Heinz made great achievements in superconducting and low-temperature experiments. Yet they were treated differently: Fritz's biography was welcomed by the *Dictionary of Scientific Biography*, but Heinz's was sent back for abridgement. Not insignificantly, the editor of the dictionary who displayed this standard preference for theory rather than experimentation was T. S. Kuhn.

Which comes first, theory or experiment?

What we have discussed above is the importance of theory and experiment in science. But which comes first, theory or experiment? Speaking also to this question, Hacking cites two great chemists as examples, albeit with quite different views: Humphry Davy and Justus von Liebig.

Humphry Davy (1778–1829) was one of the ablest chemists of his time and a teacher of the famous physicist Michael Faraday. His invention of the miner's safety lamp was a tremendous achievement. Davy wrote:

> The foundations of chemical philosophy, are observation, experiment, and analogy. By observation, facts are distinctly and minutely impressed on the mind. By analogy, similar facts are connected. By experiment, new facts are discovered; and, in the progression of knowledge, observation, guided by analogy, leads to experiment, and analogy confirmed by experiment, becomes scientific truth.
>
> (Cited in Hacking 1983, p. 152)

By contrast, Justus von Liebig (1803–1873) was a great pioneer of organic chemistry, and he advocated for the use of artificial nitrogen fertilizer, which led to an agricultural revolution. But on the question we are considering, von Liebig wrote:

> But in science all investigation is deductive or a priori. Experiment is only an aid to thought, like a calculation: the thought must always and necessarily precede it if it is to have any meaning…An experiment not preceded by theory, i.e. by an idea, bears the same relation to scientific research as a child's rattle does to music.
>
> (Cited in Hacking 1983, p. 153)

In light of these examples, and to answer the question, Hacking thus distinguishes strong and weak versions of the argument that "an experiment must be preceded by a theory." The weak version says only that we must have some ideas about nature and our apparatus before we conduct an experiment. To study nature purposelessly or without any understanding, we can achieve nothing. The strong version is that our experiment is significant only if we are testing a theory about the phenomena under scrutiny. By studying some detailed cases in the history of science and technology, Hacking himself agrees with the weak version and refutes the strong version.

Experimentation has a life of its own

Relatedly, Hacking lists seven kinds of cases to justify that experimentation has a life of its own:

(1) *Noteworthy observations*: Erasmus Bartholin (1625–1698) inadvertently discovered the double refraction in Iceland Spar, or calcite, when he examined some beautiful crystals brought back from Iceland. The phenomenon would not be explained until more than a century later by Fresnel's wave optics theory. F. M. Grimaldi (1613–1663) and then Hooke found that there was some illumination in the shadow of an opaque body; Hooke's and Newton's research on the colors of thin plates led to the interference phenomenon called "Newton's rings." The quantitative explanation of these phenomena was not made until more than a century later by Thomas Young (1773–1829) in 1802.

(2) *Wrong theories sometimes guide right experiments*: David Brewster (1791–1868) discovered biaxial double refraction, and in 1818, he published the sine and tangent laws for the intensity of reflected polarized light, which we now speak of as "Fresnel's law," five years earlier than Fresnel's wave theory. However, Brewster was not a believer in wave theory, but a thoroughgoing Newtonian! Another example is R. W. Wood (1868–1955), who, from 1900 to 1930, made great contributions to quantum optics experiments with respect to resonance radiation, fluorescence, absorption spectra, and Raman spectra. These experiments had eventually to

be explained by quantum mechanics, and yet Wood himself was skeptical of quantum mechanics. Like Brewster, Wood did great experiments not because of the correct theoretical guidance, but because of his keen observation of natural phenomena.

(3) *Meaningless phenomena*: For example, the botanist Robert Brown reported the irregular movement of pollen suspended in water in 1872. Only in the first decade of the twentieth century, due to the simultaneous work of experimenters like J. Perrin and theoreticians like Einstein, did we come to understand Brownian motion as particles being bounced around by molecules. Another example is A. C. Becquerel, who discovered in 1839 that if a pair of metal plates was immersed in a diluted acid solution, then shining a light on one of the plates changed the voltage of the cell. Yet it was not until 1905 that Einstein put forward his photon theory to explain the photoelectric effect.

(4) *Happy meetings*: In the 1930s, Karl Jansky at Bell Telephone Laboratories located a "hiss" from the center of the Milky Way. Thus, there were sources of radio energy in space, which might produce electrostatic interference. In 1965, the radio astronomers Arno Penzias and R. W. Wilson adopted a radio telescope to study the phenomenon. They detected energy sources and found that some of the energy was uniformly distributed throughout space, at around 4 K. After they had eliminated all possible noise sources, there was still a uniform temperature of 3 K. At that time, a theoretical group at Princeton University was circulating a preprint. In a quantitative way, they proposed that if the universe had originated from a big bang, there would be a uniform temperature throughout space, which would be the residual temperature of the first explosion. In addition, this energy could be detectable by means of radio signals. Penzias' and Wilson's experimental research meshed perfectly with the theory, and they won the Nobel Prize in 1978.

(5) *"Theoretician" Ampère*: The great physicist James Maxwell once praised both "inductivist" Faraday and "deductivist" Ampère. But was André-Marie Ampère (1775–1836) really a "deductivist"? Hacking's good friend and colleague at Stanford University, C. W. F. Everitt, studied this in depth, finding that Faraday's papers candidly revealed the workings of his mind, but Ampère's did not. Ampère actually paid much attention to experiments, but once he found the law and built up a perfect demonstration, he then removed all traces of the scaffolding by which he had raised it.

(6) *Invention*: Many experiments are actually inventions. For example, there were three phases of invention and several experimental concepts in the case of steam engines: Thomas Newcomen's atmospheric engine (1709–1715), James Watt's condensing engine (1767–1784), and Richard Trevithick's high-pressure engine (1798). Almost a hundred years later, in 1850, Lord Kelvin (1824–1907) coined the term "thermo-dynamic engine" to refer to all steam engines or Sadi Carnot's ideal engine, and eventually developed "thermodynamics."

(7) *A multitude of experimental laws waits for a theory*: For example, N. F. Mott and H. Jones' textbook *Theory of the Properties of Metals and Alloys* (1936) listed some experimental results and their years of discovery:

> The Wiedemann–Franz law, 1852.
> The relationships between conductivity and position in the periodic table, 1880s.
> The Matthiessen rule, 1862.
> The dependence of the resistance on temperature and on pressure, nineteenth century.
> The appearance of superconductivity, 1911.

The data were all there, but what was needed was a coordinating theory. Quantum mechanics gave qualitative explanations for almost all these experimental results in 1936, while the explanation for superconductivity came in 1957.

Pluralist relation

Do the above cases show that experiment is prior to theory? Actually, they do not. In Hacking's view, the question "Which comes first, theory or experiment?" is misleading in itself. In fact, there is no single certain answer. The relations between theory and experiment are different at different stages of scientific development, and natural sciences do not go through the same cycle. We should analyze things case by case.

The right question to ask would thus be "Which comes first, a specific theory or a specific experiment?" rather than the more general question "Which comes first, theory or experiment?" In most of the cases mentioned above, experiments came first and then theories followed. But in the history of science and technology, there are also many cases in which theory has come first and experiments have come later.

Since the relations between theories and experiments should be analyzed case by case, scientists can avoid, at least to a certain extent, the problem of circular argumentation when using experiments to support theories. For example, we used quantum mechanics to invent the electron microscope, and then it was used to test a biological theory. The observations made with the electron microscope do presuppose the theory of quantum physics, but the observation results are used to test the theory of biology, so there is no problem of circular argumentation. Here, we can reconstruct the relation between theory and experiment mentioned in the first section of this chapter as follows:

Figure 10.3 Hacking's picture of theory and experiment

Observation and experimentation

Hacking also discusses the relation between observation and experimentation. He thinks that in scientific research, observation is overrated, and we should pay much more attention to experimentation. After all, even as Francis Bacon recognized in his own time, there are actually few pure observations. Most of the time, we have to intervene in the world to get the observational results we seek.

Hacking takes microscopes as an example. We usually think that looking through a microscope is doing direct observation, but in fact, after Ernst Abbe's improvements in microscope design, even the conventional light microscope is essentially a Fourier synthesizer of first- or even second-order diffractions, thus making microscopic observation the result of intervention. Hacking used the following microscopes as his case studies:

Abbe microscope.
Polarizing microscope.
Ultraviolet microscope.
Fluorescence microscope.
Zernike phase contrast microscope.
Nomarski interference microscope.
X-ray microscope.
Acoustic microscope.

In fact, all these microscopes facilitate the collection of observation results through intervention. So in this sense, their use entails experimentation, not pure observation.

Experiments create phenomena

According to Hacking, scientists often create phenomena, which then become the core of scientific theories. When physicists study a truly instructive phenomenon by hand and mind, they call it "effect." By the second half of the nineteenth century, effects had become quite common in science, with examples being the Faraday (or magneto-optical) effect, the Compton effect, the Zeeman effect, the photoelectric effect, the anomalous Zeeman effect, and the Josephson effect.

Hacking claims that to conduct experiments is to create, produce, purify, and stabilize phenomena. To repeat an experiment seriously is to try to do the same thing better – to produce more stable phenomena with less interference.

Experimentation and scientific realism

Due to his respect for and emphasis on experiments, Hacking proposes a form of experimental realism. Since we have already mentioned his position in the previous chapter on scientific realism, here we need only quote a paragraph of Hacking's own summary:

> Experimentation has a life of its own, interacting with speculation, calculation, model building, invention and technology in numerous ways. But whereas the speculator, the calculator, and the model-builder can be anti-realist, the experimenter must be a realist. This thesis is illustrated by a detailed account of a device that produces concentrated beams of polarized electrons, used to demonstrate violations of parity in weak neutral current interactions. Electrons become tools whose reality is taken for granted. It is not thinking about the world but changing it that in the end must make us scientific realists.
>
> (Hacking 1983, pp. xiii–xiv)

According to Hacking, science is more about intervening than representing! Since we can successfully transform the world, we should believe in the reality of the world we create.

Social analysis of scientific experimentation

If Hacking's experimentalism is internal, geared toward studying experiments within science, that of Steven Shapin and Simon Schaffer is more radical. In their famous book *Leviathan and the Air-Pump: Hobbes, Boyle, and the Experimental Life* (1986), they advocate for social analysis of scientific experimentation. Our standard understanding of experiments is Robert Hooke's formality – i.e., making realistic statements in science is based on "honest hands" and "faithful eyes" – but Shapin and Schaffer, attentive to other social and even political factors, make a breakthrough in noting how:

> the establishment of matters of fact in Boyle's experimental programme utilized three technologies: material technology embedded in the construction and operation of the air-pump; literal technology by means of which the phenomena produced by the pump were made known to those who were not direct witnesses; and asocial technology that incorporated the conventions and considering knowledge-claims.
>
> (Shapin and Schaffer 1986, p. 25)

Here, we might need some background. Among his various scientific endeavors, Boyle tried to prove the existence of vacuum with an air pump. Boyle's greatest opponent at that time, however, was the British philosopher Thomas Hobbes (1588–1679). In *De Corpore* (*On the Body*, 1655), Hobbes proposed a plenist natural philosophy, in which plenist space is composed of common air, pure air, and ether. Accordingly, Hobbes denied that there can be a real vacuum in the world and thus gave another explanation for the results of Boyle's experiments. For example, for Boyle's quicksilver experiment, Hobbes believed that under strong pressure some substile substance (e.g., ether) will pass through quicksilver and all other fluid objects, just like smoke passing through water. As for the mouse that died in a vacuum in another of Boyle's experiments, Hobbes thought that this occurred because pumping brings about strong wind, which resulted in the mouse's death.

In reaction, Boyle later emphasized the following in his *New Experiments* (1660): (1) witnessing science, to perform experiments in social space; (2) prolixity and iconography for virtual witnessing, as the literary technology of virtual witnessing can produce the image of experimental scenes in a reader's mind; (3) modesty of experimental narrative, to report experimental failures; and (4) manners in dispute, as disputes should be about findings and not about persons (Shapin and Schaffer 1986, chapter 2). Following inspired lines, then, the conclusions of Shapin and Schaffer in *Leviathan and the Air-Pump* include:

> (1) that the solution to the problem of knowledge is political; it is predicated upon laying down rules and conventions of relations between men in the intellectual polity; (2) that the knowledge thus produced and authenticated becomes an element in political action in the wider polity; it is impossible that we should come to understand the nature of political action in the state without referring to the products of the intellectual polity; (3) that the contest among alternative forms of life and their characteristic forms of intellectual product depends upon the political success of the various candidates in insinuating themselves into the activities of other institutions and other interest groups. He who has the most, and the most powerful, allies wins.
>
> (Shapin and Schaffer 1986, p. 342)

Bruno Latour, a French sociologist and anthropologist famous for his research in the field of science and technology studies, has also advocated for the research methods of laboratory anthropology. From October 1975 to August 1977, Latour conducted anthropological research with Steven Woolgar at the Salk Institute for Biological Sciences in La Jolla, California. He carefully observed the daily operation in the laboratory and interviewed many experimental participants, and then Latour and Woolgar coauthored *Laboratory Life: The Social Construction of Scientific Facts* in 1979. Latour, with his colleague Michel Callon, also proposed Actor Network Theory, which marked the birth of the Paris School. The Paris School has since become an important research center for the sociology of scientific knowledge, competing with the Edinburgh School.

Summary

From a historical review of the field of philosophy of science, we can find that the mainstream of traditional philosophy of science – typified by such currents as logical empiricism, critical rationalism, and historicism – has been largely theory-oriented, paying relatively little attention to scientific experimentation in comparison. Hacking's new experimentalism advocates an experiment-oriented approach, which the author wholeheartedly applauds. The social analysis of scientific experimentation is quite popular now in historical and social studies of science, and it has great practical significance; for example, in China it has been shown to be relevant for tobacco harm reduction, the dispute over the Three Gorges Dam, the approval of the National Laboratories, and the formulation of national food standards, among other issues. These hot issues related to science and technology in China today can indeed be subjected to multi-dimensional analyses of power, culture, society, and so on.

Due to the joint efforts of new experimentalism and social studies of science and technology, the philosophical discussion of scientific experimentation has attracted more and more attention in Chinese academic circles. For instance, the Dutch philosopher Hans Radder's edited collection *The Philosophy of Scientific Experimentation* (2003) was translated into Chinese in 2012. Hopefully, these new explorations in the philosophy of scientific experimentation will enable us to have an even more comprehensive and in-depth understanding of the role of experimentation in science, both practically and theoretically.

11 Science and values

"Science and values" is by no means the biggest issue in philosophy of science today, but because it involves two such central aspects of our lives, science and values, it has inevitably aroused philosophers' lasting interest. According to Robert Hollinger, there are several important questions to answer when considering the issue: (1) Can (is/should) science be "value free" or "neutral"? This raises another question: "What is value?" because not all values are moral values. (2) If science is value neutral, what implications will it bring? If science is not value neutral, what does this imply? (3) What is the best framework for science, values, and their interaction (Klemke 1998, pp. 481–482)?

Objective values in the ancient world

In ancient Greece, there were not as many conceptual distinctions as we have today. For example, the ancient Greeks did not have the following dichotomies: (1) science and value (or knowledge and the good); (2) science and philosophy; (3) the subjective and the objective; or (4) factual or descriptive accounts of the world and normative, evaluative, or moral interpretations of the world. For Plato, "objective reality" was represented by the idea or form of "the good." The concept of morality, then, much like knowledge of geometry, had absolute objective reality. As a result, ancient Greek philosophers often deduced value judgments from facts. For example, Aristotle inferred the value judgment "Man should pursue rationality" from the fact that "Man is a rational animal." In ancient Greece, ethics were basically discussed from the perspective of facts, and ethical knowledge was regarded as objective knowledge.

Such characteristics were not unique to ancient Greece, but were also reflected in other ancient cultures. For example, in China, the Confucian philosopher Mencius (372–289 BC) also inferred the value judgment "Man should pursue benevolence and righteousness" from the following premise: "Slight is the difference between man and the brutes. The common man loses this distinguishing feature, while the gentleman retains it" (Lau 2003, p. 179). Likewise, in many ancient societies, there was almost no dichotomous distinction between science and values.

Hume's dichotomy

David Hume was perhaps the first philosopher in history to raise the dichotomy of science and value and see it as problematic. He believed that human knowledge is descriptive, with reference to what "is," whereas morality pertains to what "ought to be." These are two different kinds of category. In his *A Treatise of Human Nature*, Hume wrote:

> In every system of morality, which I have hitherto met with, I have always remark'd, that the author proceeds for some time in the ordinary way of reasoning, and establishes the being of a God, or makes observations concerning human affairs; when of a sudden I am surpriz'd to find, that instead of the usual copulations of propositions, *is*, and *is not*, I meet with no proposition that is not connected with an *ought*, or an *ought not*. This change is imperceptible; but is, however, of the last consequence. For as this *ought*, or *ought not*, expresses some new relation or affirmation, 'tis necessary that it shou'd be observ'd and explain'd; and at the same time that a reason shou'd be given, for what seems altogether inconceivable, how this new relation can be a deduction from others, which are entirely different from it. But as authors do not commonly use this precaution, I shall presume to recommend it to the reader; and am perswaded, that this small attention wou'd subvert all the vulgar systems of morality, and let us see, that the distinction of vice and virtue is not founded merely on the relations of objects, nor is perceiv'd by reason.
>
> (Hume 2007a, p. 302)

Hume's dichotomy is often called the "is/ought" (or "what-is/what-ought-to-be") dichotomy, sometimes known as the "fact/value" distinction.

Hume's dichotomy clearly divides fact and value into two categories. Facts can correspond to the objective world, so they are either true or false, and the standards of veracity and falsity correspond to the objective world. In contrast, there is no objective world to which values correspond, so values can be neither true nor false. Following this line of thinking in a certain direction, later logical positivists even advocated emotivism in the field of ethics, which is to say that all ethical judgments are ultimately our emotional expressions.

Due to this dichotomy, we now usually say that science is value neutral, or that science and morality are separated. Let us keep in mind, however, that this is not inherently a rejection of values or morality. As we might recall from above, Einstein even maintained that moral figures had contributed much more to human beings than scientists on the whole, writing in one letter:

> Humanity has every reason to place the proclaimers of high moral standards and values above the discoverers of objective truth. What humanity owes to personalities like Buddha, Moses, and Jesus ranks for me higher than all the achievements of the enquiring and constructive mind.
>
> (Einstein 2013, p. 70)

Nevertheless, in many people's minds, another dichotomy has formed: Science is objective, while moral judgments are relative, without any objective standard.

Objectivism and its criticism

With that said, in the history of modern thought, Hume's fact/value dichotomy did not come into being suddenly. It actually came from an objectivist image of modern science with respect to the world. Indeed, as early as the birth of modern science, Galileo distinguished primary qualities from secondary qualities. Primary qualities include matter, motion, and physical magnitude, which are governed by the laws of mechanics; secondary qualities, on the other hand, include color, value, interpretation, purpose, and theories, all of which are related to human subjectivity.

Rene Descartes' mind–body dualism also held that the physical realm (body) obeys the law of mechanical motion. Isaac Newton further strengthened the mechanical image of the world, with Newtonian mechanics attempting to explain all natural phenomena by means of mechanics. Later, the scope of science expanded even more to research the secondary qualities that can be reduced to the primary qualities, but it still followed the view of objectivism in the main.

Objectivism means that the world, as the subject of scientific research, is governed by laws and processes, which are independent of human beliefs, values, desires, and interpretations. Human knowledge, therefore, must correspond to the objective world. In this view, objectivity means: (1) relating to primary qualities and (2) ascertainable by means of non-subjective and unbiased methods.

Furthermore, value can be divided into "categorical value" and "instrumental value." Categorical value, also known as "intrinsic value," refers to the value that is worth pursuing in and of itself. For example, beauty is the pursuit of many girls. Instrumental value is not the goal itself, but is rather very helpful for achieving some other goals, making it also worth pursuing. For example, many people may not care about earning a degree, but a good degree can bring more employment opportunities and better economic remuneration, making it still worth pursuing. The distinction between categorical value and instrumental value is only relative. For example, many people regard physical health itself as an intrinsic value worth pursuing, but some may think that physical health is important primarily for career success, so it is only an instrumental value.

According to objectivism, value judgments can be expressed either in scientific language or in pure subjective expressions of personal preference. Therefore, objectivists have two attitudes toward value judgments. One is that all categorical values are descriptions of the objective world; for example, naturalists believe that "the good" means "pleasure," so value judgments can be reduced to some natural conditions of human beings. The opposite view is that all value judgments are arbitrary, so they are irrational or non-rational. If science contains value judgments, it is only ideology in the end.

German philosopher Edmund Husserl (1859–1938) thought that these two attitudes share the same thing – or, in Kuhn's language, "paradigm": (1) Everything can be divided into objective and subjective; (2) only scientific knowledge is objective; (3) ultimate value judgments are either scientific or subjective; and (4) ultimate values cannot be rationally debated. Therefore, both of them presuppose the logic of objectivism.

Objectivism brings two clusters of ideas: (1) Many either turn ethics into science by trying to explain value judgments scientifically, such as historical or economic determinism, or fall into moral nihilism or extreme ethical relativism; (2) some try to reconcile "objective" science and "subjective" morality, and thus save human morality and freedom by determining a limit to science. They believe that science and morality do not conflict, but rather complement each other; they have nothing to do with each other, but instead study different spheres of experience – i.e., "man as object" and "man as actor."

Objectivism also led to instrumental rationality or technical rationality. According to the logic of objectivism, rational justification is essentially hypothetico-deductive because the method is logical and objective. Max Weber (1864–1920), a German thinker and sociologist, thus developed the concepts of "instrumental rationality" and "technological rationality." Weber believed that the world is meaningless and in infinite flux, so we must take a specific point of view or framework, which is determined by either our personal interests and values or those of our cultures. Weber's view is also called the thesis of "value relevance" (Hollinger 1998, pp. 539–549).

In Weber's view, values are subjective because they cannot be proven by science. Science is value neutral, as it can only tell us what the consequences of our subjective decisions are, so that we can know the consequences of actions and clarify the responsibilities. Weber called this an "ethics of responsibility."

Thus, Weber's view of rationality is actually instrumental rationality following the method of cost-benefit analysis. For example, my quitting smoking is very effective for the goal of my physical health, so "quitting smoking" is rational for my "physical health"; physical health can make me better engaged in philosophical research, which is effective for my goal of becoming a philosopher, so "physical health" is rational for "becoming a philosopher." I want to be a philosopher because my parents will be happy with me, so "becoming a philosopher" is rational for "making parents happy." But there must be an end to the chain; for example, "making parents happy" cannot be instrumental for "quitting smoking." Therefore, for ultimate values, we can no longer justify rationality; they are simply "radical choices." Instrumental rationality basically follows such a layered structure:

Figure 11.1 Weber's picture of instrumental rationality

In this way, Weber contends that ultimate values can only be radical choices, which are irrational. Hume took a similar view, famously saying: "Reason is, and ought only to be, the slave of the passions."

Weber pointed out that our biggest problem in modern society is that there is no objective value in our lives, so human beings have lost common goals for which to strive. In addition, science and technology have benefited human beings greatly in recent years, but the disasters brought about by technological development are also increasingly prominent, so the saying "Science and technology is a double-edged sword" has become increasingly common. Even as this broad example shows, while rethinking the social functions of science, people have begun to reflect more deeply on the objectivism underlying modern science.

Consequently, Robert Hollinger summarizes some refutations of objectivism: (1) In the movie *Love and Death*, Woody Allen says, "Objectivity is subjective, and subjectivity is objective." He uses the sentence to show that people's subjective feelings can sometimes be as objective as other facts. For example, when we say, "I am in pain," our subjective feelings about pain, like our weight, are objective. (2) Objective knowledge comes from the "rational reconstruction" of private and subjective experience, so modern science is rooted in subjective experience. (3) Michael Polanyi, a historian and philosopher of science, proposes that knowledge, truth, and objectivity are all rooted in human values and human purposes. Otherwise, science need not study the world in which human beings live because the human world has no significance compared with the universe. (4) The pursuit of knowledge itself embodies a kind of value. In addition, the distinction between reliable and unreliable knowledge, good and bad methods, also needs normative value judgment (Klemke 1998, pp. 484–485).

Values in science

Rudner: The scientist qua *scientist makes value judgments*

In 1953, Richard Rudner claimed that science needs values, in a paper titled "The Scientist *Qua* Scientist Makes Value Judgments." Rudner pointed out that most defenses of the values in science in the past appealed to three points: (1) Our having science already involved a value judgment; (2) scientists must make value judgments when choosing alternative scientific problems; (3) scientists are also human beings, who are "masses of predilections," which will also affect their scientific activities. To these three points, some may argue that the first two value judgments are actually extrascientific or pre-scientific, not internal to science. As for the third point, the perfect scientist, or scientist *qua* scientist, will not let their value judgments affect their research.

Rudner makes clear that the scientist *qua* scientist also makes value judgments, however, because when scientists engage in scientific research, they often need to accept or reject scientific hypotheses. Since scientists have to decide whether the evidence is strong enough, they need to judge importance, which is an ethical judgment in some sense. In addition, the degree of certainty with which we accept hypotheses often depends on how serious a mistake might be, so scientists need to make value judgments before accepting or rejecting a certain hypothesis. Rudner writes:

> Any adequate analysis or…rational reconstruction of the method of science must comprise the statement that scientist qua scientist accepts or rejects hypotheses; and further that an analysis of that statement would reveal it to entail that the scientist qua scientist makes value judgements.
> (Rudner 1998, p. 495)

Following a different tack, Rudolf Carnap proposed a dichotomy between external and internal questions in science. The external question is used to decide which theoretical language or framework we should choose (for example, Euclidean geometry or non-Euclidean geometry in the general theory of relativity), while the internal question is used to discuss within the theoretical framework (such as the sum of the extension angles of triangles in Euclidean geometry). Carnap believed that external questions are practical ones that involve value questions, while internal questions are theoretical ones, which can be derived from the basic axioms and syntactic inferences within a linguistic framework.

Rudner, however, inherited W. V. Quine's tradition, which argues that science is a unified structure, so there is no dichotomy between external and internal questions; similarly, theory choice will spread to every scientific hypothesis. According to Rudner, then, all scientific theory choices ultimately entail value judgment. Such a view that theory choice needs value judgment will not necessarily render scientific methods out of control, though, as Rudner advocates a new science of ethics as well, studying the value judgments operative in science and aiding science's progress toward objectivity.

Hempel: Scientific knowledge needs valuational presuppositions

C. G. Hempel pointed out in his 1960 paper "Science and Human Values" that scientific activities presuppose valuations: (1) When scientists choose to engage in scientific research rather than other activities, they have already presupposed the value judgment that scientific research is more important; (2) scientists' selection of topics also presupposes values. For example, scientists choose dust storm prevention and control, rather than dust storm diffusion, as the research project, which also presupposes a value choice.

More importantly, Hempel pointed out that scientific theory choice demands some kind of value presupposition. The problem of induction and the thesis of underdetermination have shown that limited observations cannot logically determine which scientific theory is correct, so scientists need to choose among infinite theories in principle. In a theory decision, three aspects are required in the decision making: (1) factual information – e.g., what kind of consequences different choices will bring; (2) utility assignment, giving different utilities to different consequences; and (3) selection criteria, such as the more conservative principle of minimum loss or the more aggressive principle of maximum benefit. There is no doubt that the latter two aspects need value judgments. Therefore, scientists need to make value judgments when engaging in theory choice. Since theory choice is very common in scientific research, science in fact presupposes values.

From this, Hempel's conclusion was that science cannot provide valid proof for value judgments, but scientific knowledge needs to presuppose values. At the same time, he also acknowledged that science in turn plays important roles in defining values: (1) Science can provide factual information for value judgments, such as whether a contemplated objective can be attained in a given situation, which method has the highest probability of achieving the objective, what side effects or final consequences the choice will bring, and whether several ends can be realized jointly or are incompatible. (2) Some research results of psychology and sociology may affect values or contribute to emotional security. For example, Darwin's evolutionism led to social Darwinism, and Freud's sexual theory also had an outsized impact on the sexual revolution. (3) The development of science can also give us some insight. Science is constantly developing, and no matter how grand or deep a theory is, it will eventually be replaced by a new theory. This also inspires us to see how perhaps ethical judgment should also be "relatively ultimate," meaning that to move forward we will, more than ever before, need undogmatic, critical, and open minds (Hempel 1998, pp. 511–514).

Kuhn: Value judgment and theory choice

In his 1973 paper "Objectivity, Value Judgment, and Theory Choice," Kuhn also argued that scientific theory choice is neither subjective nor objective, but based on the "collective judgment" of scientists trained in the community.

According to Kuhn, there is no objective rule or proof for theory choice. Scientists' decision to accept or abandon a given theory or paradigm "cannot be resolved by proof." The mechanism involves techniques of persuasion, or argument and counterargument, in the situation without proof. Therefore, theory choice is not purely objective; rather decisions are made by scientists according to certain values, like accuracy, consistency, broad scope, simplicity, and fruitfulness. Accuracy means that the consequences derivable from a theory should be in demonstrable agreement with the results of existing observations and experiments. Consistency requires that a theory not only be consistent internally with itself, but also with currently accepted theories. Broad scope means that a theory's consequences go far beyond the specific observations or laws that it was originally designed to explain. Simplicity requires that a theory be concise in mathematical form and avoid increasing unnecessary presuppositions. Fruitfulness means that a theory produces many new research results. Such value judgments provide a common basis for theoretical choice.

Kuhn believed that theory choice depends on scientists' value judgments, so it is neither objective nor algorithmic. Nevertheless, he denied that theory choice thus becomes subjective and arbitrary or descends into mob psychology. He proposes that the word "subjective" has at least two usages: one is opposed to "objective" – that is, including humans' subjective factors; the other is opposed to "judgmental" – that is, a problem of a matter of taste that cannot be discussed. Upholding that theory choice is not algorithmic and logical, Kuhn used the word "subjective" in the opposite sense of "objective." However, many of his critics have confused the two usages of "subjective," thinking that Kuhn made theory choice a matter of taste that cannot be discussed, and thus seriously misunderstanding Kuhn's position. After all, Kuhn claimed that objectivity should be analyzable in terms of criteria such as accuracy and consistency, writing: "If these criteria do not supply all the guidance that we have customarily expected of them, then it may be the meaning rather than the limits of objectivity that my argument shows" (Kuhn 1977, p. 338). Obviously, Kuhn's conclusion provides a new way to break the dichotomy of science and values, as well as that of objective and subjective.

Many subsequent new schools in philosophy of science have further developed Kuhn's thought. For example, feminists suggest that science often presupposes male values such as governance and control, making it biased. The sociology of scientific knowledge also questions the concepts of truth, knowledge, reality, objectivity, and rationality, and holds that science is only the result of social construction – no more objective than stories, novels, and folk tales. Some postmodernist philosophers propose the concept of "blurring of the genres," believing that physics and history are merely different forms of text or writing.

Summary

Though the issue of science and values is not the most central in philosophy of science, it reflects on the nature of science and its position in human life from the outside of science. Because of this, if there is any great progress on this issue, it may lead to a revolution in human thought. The work of German philosopher Jürgen Habermas (1929–) is worth noting in this respect.

Habermas criticizes the traditional view, represented by Weber's instrumental rationality, as a mixture of scientism and decisionism. Scientism, so called, means that only the methods and results of natural science are effective and rational, so there is only one kind of knowledge – namely scientific knowledge – and only one kind of rationality – namely technological rationality. Decisionism means that ultimate value judgments cannot be reduced to scientific knowledge, nor can they be proven by science or scientific methods. Therefore, value judgments only express individual and arbitrary decisions. As Habermas sees it, scientism and decisionism are actually two sides of the same coin.

In his book *Knowledge and Human Interests*, Habermas points out that human knowledge is rooted in universal human interests, of which there are three kinds: technical interest, practical interest, and emancipatory interest. Technical interest relates to human beings' need to control nature for survival. "Labor" is such an activity, and modern science and cost-benefit rationality best serve our technical interest. Practical interest is that human beings want communication, interaction, and a public life. Related knowledge must be hermeneutic, or interpretive, so as to expand the capacity of human communication. Emancipatory interest aims to liberate human beings from oppressive forces, including material, political, psychological, and ideological forces. Critical theory is thought to best further this goal.

Accordingly, Habermas believes that, relative to these three human interests, there should also be three types of human knowledge – natural science, human science, and critical theory – and these three types of knowledge should not be replaced or confused with one other. The mistake of objectivism is to extend natural science to all areas of human knowledge.

If Habermas' classification of interest and knowledge is established, then the nature of natural science and its position in human knowledge will be clear. Regardless, as this potential shows, the issue of science and values is of great significance for us to understand the nature of natural science and to reflect on our own image of the world.

12 New developments in philosophy of science

The changes of textbooks in philosophy of science

Thomas Kuhn once said that one quick way to find the center of a certain field is to look at its textbooks. Therefore, if we want to understand the new developments in philosophy of science, we had better look at the changes of textbooks. From this we can hopefully gain some inspiration.

In the 1960s and 1970s, the main textbook of philosophy of science was Ernest Nagel's *The Structure of Science: Problems in the Logic of Scientific Explanation* (1961). This book contains a very detailed introduction to logical empiricism. Another major textbook during this period was *The Structure of Scientific Theories* (1969), edited by Frederick Suppe. This is a collection of papers from the philosophy of science symposium in Urbana, Illinois on March 26–29, 1969, and its influence was so pronounced that at the 1998 biennial meeting of the Philosophy of Science Association, a special symposium called "The Structure of Scientific Theories: Thirty Years On" was organized to commemorate the book. On the whole, the textbooks from that period mainly analyze the logical structure of scientific theory from within science. Therefore, static analysis was dominant, as can be seen from the emphasis on "structure" in the titles of the two textbooks. If students majoring in philosophy of science want to know more about such developments, they can also read Thomas Kuhn's *The Structure of the Scientific Revolutions* published in 1962, and Israel Scheffler's critical book *Science and Subjectivity* (1967).

Twenty years later, the focus of philosophy of science was no longer logical positivism, as more attention shifted to the major figures David Stove has called the "four modern irrationalists": Karl Popper, Thomas Kuhn, Imre Lakatos, and Paul Feyerabend. In fact, Popper and Lakatos emphasized the rationality of science, making them "critical rationalists," but because of their disagreement with logical positivism, they were also included in the camp of "irrationalists," along with Kuhn and Feyerabend.

In the 1990s, new textbooks mainly included *The Philosophy of Science*, (1991) edited by Richard Boyd et al., and *Introduction to Philosophy of Science*, (1992) edited by Merrilee Salmon and her colleagues at the Department of History and Philosophy of Science, the University of Pittsburgh. The late 1990s saw the release of the third edition of *Introductory Readings in the Philosophy of Science* (1998) edited by E. D. Klemke et al., a text for beginners in philosophy of science, which covered the relatively major topics and provided selected readings from classic texts in the field. *Philosophy of Science: The Central Issues* (1998), edited by J. A. Cover and Martin Curd, also came out during this period. Both of these books introduce the main issues of philosophy of science in a more traditional way.

In contrast to these traditional introductions, a number of more radical textbooks were also published in this period. For example, George Couvalis published *The Philosophy of Science: Science and Objectivity* (1997), which responds to Scheffler's criticism of Kuhn in *Science and Subjectivity* and includes discussions on relativism, sociology of scientific knowledge, and feminism. Additionally, Robert Klee's edited volume *Scientific Inquiry: Readings in Philosophy of Science* (1999) discusses social constructivism and feminism in the second part, titled "Historicism and Its Aftermath." Furthermore, in *Scientific Knowledge: Basic Issues in the Philosophy of Science* (1998), Janet Kourany not only analyzes the social situation of scientific knowledge, but also deals with some traditional issues in philosophy of science from the perspectives of feminism and social constructivism. Finally, Jennifer McErlean's *Philosophies of Science: From Foundations to Contemporary Issues* (2000) introduces postmodern rhetoric, social critiques of science, narrative, metaphor, and feminism.

From the changes of textbooks, we can roughly see that the field of philosophy of science has gradually changed from static analysis to dynamic research, and from internal to external studies. Why do the textbooks in philosophy of science express such changes? If such a change is not arbitrary, it is probably because the traditional philosophy of science contained some problems that needed changing.

Famed philosopher Alasdair MacIntyre advocated adopting Hegel's narrative approach to philosophical research. Such a narrative approach uses the method of storytelling to outline the history of academic development, so as to understand the discipline. Seen this way, philosophy of science can be traced back to the controversy between rationalism and empiricism of modern philosophy, to Comte's positivism, or, for some philosophers, even all the way back to ancient Greece. Nevertheless, it is generally held that philosophy of science as a discipline began with logical positivism, and logical positivists' logical analysis of science is regarded as a typical static approach to philosophy of science. In light of this, the transformation of philosophy of science from static analysis to dynamic research may be due to the problems in the static analysis of logical positivists, the solution to which was enabled by dynamic research, just as it has paved the way for more such research going forward. But what exactly is the problem with logical positivism, anyway?

The sin of logical positivism and relativism

Logical positivism was the philosophy of the Vienna Circle in the 1920s and 1930s, proposed as a "scientific philosophy" opposed to the more traditional philosophy of "metaphysics." Starting from this point, research interests focused more on science itself, and finally "philosophy of science" was established. Most members of the Vienna Circle were scientists or had advanced scientific training, and thus had a deep understanding of new developments in physics at that time, such as the theory of relativity and quantum mechanics. Largely because of this, their philosophical views were not only dominant in philosophy, but they were also quite popular among scientists themselves – even coming to be called "standard philosophy," or the "received view."

Obviously, however, logical analysis alone is not enough to deal with all the problems confronted in philosophy of science, and many logical difficulties were encountered. Because logical positivists used explicit methods to express ideas clearly, their mistakes were also highlighted in a very obvious form. Picking up on these mistakes, Quine criticized what he perceived as two dogmas of empiricism: (1) the dichotomy of analytic–synthetic propositions, or that the truth value of analytic propositions comes from meaning, having nothing to do with facts, and synthetic truth comes from empirical fact; and (2) reductionism, which holds that each meaningful proposition is equivalent to the logical construction of some terms, and that these terms can point to direct experience, which is to say that meaningful statements can be reduced to direct experience (Quine 1980, 20–46). Though figures like Carnap and Strawson tried to defend the dichotomy of the analytic–synthetic, Quine's criticism nonetheless shook the very foundations of logical positivism.

Logical positivists' meaning criterion was also challenged. Because of apparent logical problems, the meaning criterion was changed from the principle of verification to the principle of falsification, and finally relaxed again to the principle of confirmation. However, Church's formula made even the principle of confirmation difficult to achieve. As a result, the meaning criterion developed from the principle of testability to the principle of translatability, which eventually led to the rise of holism and finally caused the abandonment of any sharp distinction between meaningfulness and meaninglessness.

Logical positivism also holds that there is a neutral observational language that can determine the truth of scientific theories. But according to the thesis of underdetermination, there can be, in principle, infinite possible theories to correspond to limited observation results. The dichotomy of observation–theory was also criticized by Karl Popper, who felt that observation is theory-laden, meaning there is no neutral observation prior to theory. However, Popper still maintained the account of a crucial experiment that can neutrally judge truth and falsity between competing scientific hypotheses. The pendulum shifted even more when Kuhn's normal science criticized Popper's account of the crucial experiment, and when Popper's successor Lakatos also proposed a sophisticated falsificationism that denied the crucial experiment.

Whether there is a neutral observational language is also related to the concept of "translatability." According to this thesis, competing theories can be translated into neutral observational language, which can be used to test the truth of the theories; or the old theories can be completely translated into the new theory language, which can fully explain the old theories – i.e., the old theories can be regarded as some limiting cases of the new theories. However, the thesis of translatability was disputed by Quine's position of the "indeterminacy of translation" (Quine 1998, 259–266). Kuhn's concept of incommensurability, moreover, directly refuted the thesis of translatability.

Logical positivists advocated using logical language or artificial language as the universal language to analyze and normalize natural languages, and, in so doing, attributed the fallacies of traditional philosophy to the illogical use of language. However, the view that logical language is a universal or normative language was even abandoned by Wittgenstein himself, as later in life he put forward the concept of "language games" and argued that logical language is only one of many language games and cannot cover all natural languages.

Logical positivists wished to use scientific methods to explain the rationality of science: Those conforming to scientific methods are rational, and those not conforming to scientific methods are irrational. Carnap tried to develop inductive logic as the scientific method: For a given bit of evidence, inductive logic can calculate the supporting degree of evidence for a particular theory. But Goodman's grue paradox argued that the method of inductive logic based on degree of confirmation could not succeed. Popper denied the role of induction in science and admitted that there was no logical path for scientific discovery, even if his hypothesis falsification method can still be regarded as a method of scientific justification. The Duhem–Quine thesis then denied that any single hypothesis can be falsified by observations. But without the scientific method, how can we explain that science is rational?

The theoretical difficulties of logical positivism thus opened the door to post-positivism, especially relativism. For this reason, Larry Laudan called logical positivism the "sins of the fathers" of relativism (Laudan 1996, p. 3).

Kuhn's great book *The Structure of Scientific Revolutions* came out in 1962 within this context. Though Popper claimed that it was he who had "killed logical positivism," it is widely believed that the publication of *The Structure of Scientific Revolutions* brought about the "paradigm shift" in philosophy of science. This representative work of historicism initially discusses "normal science" and "scientific revolution" from the perspective of the history of science. However, the times produce their heroes. Its specific historical background made this book a turning point in the criticism of traditional philosophy of science. Ironically, the book was written for the *International Encyclopedia of Unified Science* at the invitation of Carnap, a representative of logical positivism, yet the publication of the book announced the end of logical positivism.

In the book, Kuhn put forward the concept of a "paradigm." When he used this concept, however, his definition was rather vague. In fact, Margaret Masterman summed up 21 different usages of "paradigm" found within its pages. In response, Kuhn later preferred the concept of a "disciplinary matrix" to express the same meaning. A disciplinary matrix includes: (1) symbolic generalizations – i.e., scientific concepts or terms; (2) shared beliefs, which include metaphysical worldviews or theoretical models; (3) shared values, which the scientific community has cultivated as a sort of connoisseurship of scientists; and (4) exemplars.

Furthermore, Kuhn believed that the development of science is the result of the alternation of normal science, which is engaged in the activity of "puzzle-solving" within a paradigm, and scientific revolution when a paradigm shifts. He further noted that there is incommensurability between paradigms. "Incommensurability" means that there is no common basis between two paradigms for rational comparison, and it can be analyzed in three aspects: (1) different scientific criteria or norms; (2) conceptual change; and (3) different worldviews.

The concepts of "paradigm" and "incommensurability" raised severe challenges to the rationality and objectivity of science. Although logical positivists usually preferred intersubjectivity to objectivity and readily regarded the debate over realism as metaphysics, the logical positivist view is in fact closer to realism. For example, Schlick was a critical realist in his early stage and later turned to empirical and positivist realism. Carnap and Neurath also had strong realist tendencies. According to logical positivism, the process of scientific development is the continuous accumulation of scientific theories. In fact, it presupposes that scientific theories are real descriptions of our empirical world, so the development of science is the continuous accumulation of real descriptions of the empirical world.

Although Popper replaced the accumulation theory of logical positivism with his theory of continuous revolutions, he still retained the concept of "verisimilitude" – i.e., that a new theory is closer to the truth than the old one. Seen this way, therefore, science can still be regarded as an objective description of the real world. His "three worlds" theory fully expressed such a belief in "objective knowledge."

But Kuhn's concept of a paradigm also brought real challenges to realism. If the development of science includes normal science within a paradigm and paradigm shifts, and there is incommensurability between paradigms, then how can we claim that only a new theory is the objective description of reality? Can the old paradigm not be closer to reality? For example, the atomism of ancient Greece was displaced by Plato's and Aristotle's theories for centuries, but it was revived by modern chemistry at the beginning of the nineteenth century. Who can guarantee that our scientific paradigm today best expresses objective reality and will not be replaced by a new paradigm in the future?

The rationality of science became problematic under the impact of the concepts of paradigm and incommensurability. Logical positivists used the "scientific method," such as inductive logic, to explain "scientific rationality," but they failed. According to the viewpoint of instrumental rationality, a more reasonable way would be to find a universal standard, such as "knowledge growth" or "predictive power," and then measure the degree of rationality by comparing its effectiveness.

Such a universal standard is possible within a paradigm because the paradigm defines the common goal and evaluation standards of the scientific community. However, the concept of incommensurability denies the possibility of rational comparison across paradigms. First of all, the incommensurability of languages hinders communication between paradigms. Next, the incommensurability of evaluation standards also denies that different paradigms have the same standards or values, but may be different according to their respective criteria. Finally, if different paradigms have different worldviews, but there is no common worldview as an absolute frame of reference, then all paradigms or worldviews have the same claim upon reality and we cannot make rational choices. Ergo, Ernan McMullin, former chair of the Department of History and Philosophy of Science at the University of Notre Dame, acknowledged that, "Thirty years later, *The Structure of the Scientific Revolutions* still leaves us an agenda" on the issue of rationality (McMullin 1993, p. 76).

Kuhn's challenge to the objectivity and rationality of science influenced the rise of relativism in philosophy of science. Relativists hold that there is no absolute objectivity and rationality in science, as both objectivity and rationality are relative to a paradigm, worldview, culture, tradition, discourse, perspective, etc. Therefore, different paradigms are not comparable, have the same rights, and are equally good. The problem of relativism has also opened the way for the "new age" of philosophy of science.

"New age" philosophies of science

In her paper "'New Age' Philosophies of Science: Constructivism, Feminism and Postmodernism," Noretta Koertge called constructivism, feminism, and postmodernism the "new age" philosophies of science. She was a professor of history and philosophy of science at the University of Indiana and the chief editor of *Philosophy of Science* in 2000, and her review of the latest developments in philosophy of science is quite convincing. In fact, all three trends that she mentioned are closely related to relativism.

Constructivism

Constructivism can also be called "social constructivism." Steve Woolgar and Malcolm Ashmore define constructivism as follows: "Scientific and technical knowledge is not the rational/logical exploration from existing knowledge, but the contingent product of various social, cultural and historical processes" (Woolgar and Ashmore 1988, p. 1). Constructivists regard scientific knowledge as the result of scientists' consensus, making it a contingent product, and then use sociological methods to study how scientists reach consensus. Their research in sociology is also known as "sociology of scientific knowledge" (SSK).

SSK originated at the University of Edinburgh. In the early 1970s, scholars like Barry Barnes, David Bloor, Steven Shapin, and Andrew Pickering proposed that scientific knowledge is a social product of institutionalized scientific research, so scientific knowledge should also be the object of sociological research. In order to distinguish it from the orthodox sociology of science of the Mertonian School, they named their own sociology "sociology of scientific knowledge." Harry Collins at Bath University, Bruno Latour, and Steve Woolgar are also members of this SSK school (Liu Hua-jie 2000, pp. 38–44).

A great deal of specific case studies have been conducted under the aegis of SSK to show that the production of scientific knowledge is the result of negotiation between scientists. For example, Andrew Pickering put forward in *Constructing Quarks* that the emergence of quark theory was not a scientific discovery, but the result of those in the scientific community who believe in quarks finally defeating the scientists who were against quarks (Pickering 1984). Donald MacKenzie also said in his article "Statistical Theory and Social Interest" that some scientific achievements of eugenics were accepted because the efficiency of a scientific school is better than that of the opposite school, so it attracts scientific attention to its own work (MacKenzie 1978, pp. 35–83). David Hull's *Science as a Process* and Philip Kitcher's *The Advancement of Science* study the impact of the scientific reward system and standards of professional expertise on the production of knowledge.

In *Knowledge and Social Imagery*, David Bloor introduces four principles for the "Strong Programme" of SSK (Bloor 1976):

(1) *Causality*: Social or other types of causes lead to the formation of beliefs or knowledge.
(2) *Impartiality*: We should treat truth or falsehood, rationality or irrationality, success or failure, impartially.
(3) *Symmetry*: The same cause should be able to explain a true or false belief at the same time.
(4) *Reflexivity*: The explanation model of SSK can be applied to itself in principle.

Among these, however, SSK's principles of "impartiality" and "symmetry" have aroused great disagreement among scientists. Some have particularly been disgusted with SSK's denial of the objectivity of scientific knowledge, instead preferring Robert K. Merton's or Talcott Parsons' traditional sociology of science. Accordingly, many scholars have also criticized constructivism from the perspective of case studies. For example, A. Franklin has criticized Pickering's theory of constructing quarks from the perspective of the history of physics (Franklin 1998), and P. Sullivan has also criticized Mackenzie's case study (Sullivan 1998). Against constructivism, Ian Hacking provides a systematic analysis of constructivism in *The Social Construction of What?* (1999). In *The Construction of Social Reality* (1995), John Searle insists that social reality is built on the basis of brute facts that guarantee the position of realism and anti-relativism.

If, as constructivists claim, there are social factors in scientific research, then science is not purely rational or objective. Indeed, nowadays, it is difficult for even traditional philosophers of science to deny that there are many social factors in scientific research. The research of SSK has also made progress in this respect. Many SSK scholars conduct field studies in laboratories and have a deep understanding of the daily activities of scientific research. The traditional philosophy of science holds that science has static logic beyond historical contingency, and rarely confronts the role of irrational or social factors in scientific research. SSK thus places science against a broader social background, undoubtedly enriching our understanding of science in the process. Such critical research on scientific systems is also conducive to the healthy development of science.

Nevertheless, constructivists further contend that the social system of science will directly affect the correctness of scientific research, and this remains a point of controversy between constructivists and scientists. The main difference between constructivism and anti-constructivism is the view as to whether scientific knowledge is objective or constructed by the scientific community. Constructivists hold that so-called scientific knowledge is comprised of beliefs accepted by the scientific community; therefore, the establishment of scientific knowledge is the result of a victory of the contingent of the scientific community that believes in a given theory over and against the opposition. But the result, in this view, is not objective and inevitable; it is contingent and social. If the opposing faction has more members, is more efficient, and has the upper hand in the debate, then scientific knowledge will be reversed.

By contrast, anti-constructivists believe that social factors in scientific research will not affect the objectivity of scientific knowledge. Science is more natural than social. A scientist may be engaged in scientific research for position or money, and the social system of science may hinder the development of science at some times, but this does not affect the objectivity of scientific knowledge.

The author is of the view that the constructivists' inference presupposes the idea that "because scientists believe, scientific knowledge becomes true." Accepting this, in order to obtain scientific knowledge, we should study "how scientists believe" from the perspective of sociology. While anti-constructivists' presupposition is that "because scientific knowledge is true, scientists believe it," the truth of scientific knowledge is not affected by social factors. But the problem is how we can know whether scientific knowledge is real. If scientific laws are objective, then scientists can only "discover" scientific knowledge, not "construct" it. However, Kuhn's concept of a paradigm completely shakes the presumption of the objectivity of science: Since different paradigms may entail different worldviews, and those paradigms are incommensurable, who can guarantee that the current paradigm is objective and will not be replaced by a new paradigm in the future? In addition, Kuhn's concept of "scientific community" also makes people realize that scientific research is not an activity of individual scientists, but is rather a social enterprise of the whole community. This makes SSK, as a sociological study of the scientific community, legitimate.

As we have seen, the rise of constructivism was closely related to Kuhn's criticism of traditional philosophy of science. Likewise, the ultimate overcoming of constructivism lies in how to explain the "objectivity" of science again after Kuhn.

Feminism

In the West, feminism, as a political movement, was initially devoted to striving for women's equal rights with men economically and politically. Later, it developed an academic aspect, which provides a female perspective on how to view the world. This part deals with feminism in the academic sense.

But first, let us clarify some terminology in feminism. "Woman" refers to an individual, such as "Madame Curie is a woman." "Sex" applies to biological difference. "Gender" refers to differences associated with social or cultural factors. Biological sex includes male and female, while gender is described as "masculine" and "feminine" in the social and cultural characteristics of men and women.

Early feminists ignored the difference between men and women and thought that women were the same as men in thought and action. They desired for women to be like men culturally – and, at times, even physically. In their attitude toward science, the hope was that more women would join the ranks of scientists, but not that science itself needed to change. For example, early feminist works, such as M. Rossiter's *Women Scientists in America* and H. Zukerman's *The Outer Circle*, both maintained that science is good and that it is a shame not to have more women in science. The goal of their research was thus to expand the role of women in science as it was already practiced. Londa Schiebinger called this early feminism "liberal feminism."

However, there are obvious differences between men and women, both physically and socially, and later feminists have emphasized these differences. Schiebinger called this approach "difference feminism" in *Has Feminism Changed Science?*, and C. Sommers called it "gender feminism" in *Who Stole Feminism?* Difference feminists' main viewpoints include: emphasizing the differences between men and women, rather than the similarities; valuing and evaluating the feminine values that were regarded negatively in traditional societies, such as subjectivity, cooperation, emotion, empathy, etc.; and attaining women's equality in science not only by changing women's access and status, but also by changing science itself, from curriculum, to laboratory, to theory, to research programme, and so on. Proponents of this approach may tend to think that women have a unique way of knowing. Although this cognitive style is suppressed by male scientists, who are currently dominant in science, it helps female scientists to achieve outstanding results, and may even change the face of science in turn. An example of this would be Sue V. Rosser's call for "female-friendly science" (Rosser 1990).

In another example, Evelyn F. Keller has made a case study of Barbara McLintock, a geneticist who won the Nobel Prize in 1983. Keller believes that McLintock's outstanding achievements in the field of science are due to her different emphasis on "feelings for the organism," rather than the domination of nature, as most male scientists might emphasize. In McLintock's view, the subject and object, the observer and the observed, are no longer completely separated. We must care about and empathize with the research object. However, McLintock's research method was suppressed by the male-dominated scientific community, and her research was not officially recognized with the Nobel Prize until 30 years later.

Schiebinger has also listed feminists' contributions to science. For example, in medicine, the National Institute of Health established the Office of Research on Women's Health in 1990, and Women's Health Action has provided funding for neglected research on women's diseases since the 1990s. Anthropologists and archaeologists have revised the account of "the first tools" to recognize the role of women in human evolution. Some progress has also been made in biology; for instance, in primatology, male scientists used to study only in the laboratory, but later, feminist scientists went deep into the jungle, lived with primates, and collected rich materials. Their research has also shown that female primates not only exchange food through sexual behavior, but also play important roles in the population. Because of the remarkable achievements of women in this field, primatology was once called a "feminist science." Even still, Schiebinger acknowledged that feminism, at least by the time of publication, had made little progress in the fields of the physical sciences and engineering sciences (Schiebinger 1999).

With that said, difference feminism has also been challenged on various points. Donna Haraway and Judith Butler have raised the critique that feminists have too easily presupposed the concept of a "universal woman" and believed that women have certain common characteristics, when in fact women often show great differences due to background in terms of country, race, culture, and class. The concept of "women's way of knowing" has been criticized as problematic, too. For example, Katharine Hayles has contended that holism is not a unique way of knowing for women. In fact, non-linear system research, such as chaos theory, has developed rapidly in recent years, but few women scientists have been involved in the research.

As a result, some feminists have become more moderate. For example, Schiebinger denied that feminine values such as cooperation and empathy are unique to women, as male scientists can also express these. However, she stressed that gender research can help the scientific community to further enrich its methods and contents, and that feminism can play a significant role in this regard.

In fact, the development of feminism is closely related to Kuhn's impact on philosophy of science. According to the traditional philosophy of science, natural science is the only objective description of the world. At most, then, feminism can only enrich the theme of scientific research, such as through concern for women's diseases, but it will not affect the objectivity of scientific content. For this reason, the role of feminism will not be highlighted. Kuhn's historicism, however, denies that human beings can have an objective and unique description of the world, as there is no objective "seeing," but only different "seeing as," depending on various perspectives. These perspectives are "incommensurable" and have the same rights. There is no right or wrong, just different perspectives. As a result, because feminism provides a new perspective on the world, it should have the same rights as the purportedly male perspective provided by traditional science. Feminism can even be used as a supplement to the male perspective, thus profoundly changing the future of science.

In this sense, feminism has gained wide support and sympathy, as feminist perspectives may bring new and different perspectives to science. Even if feminism turns out to be wrong on certain points, there will be no social harm, so it is of course right to do such a good and harmless thing.

Postmodernism

As the name suggests, postmodernism is a criticism or surpassing of "modernity." If the term "modern society" refers to the whole of Western society since the Renaissance, the content of "modernity" can include many aspects, such as the capitalist economic system, the liberal political system, etc. The philosophical dichotomy of subject and object, instrumental rationality, scientific hegemony, and technological control can also be attributed to the phenomenon of "modernity."

Jean-François Lyotard uses the term "modernity" as follows:

> to designate any science that legitimates itself with reference to a metadiscourse…making an explicit appeal to some grand narrative, such as the dialectics of Spirit, the hermeneutics of meaning, the emancipation of the rational or working subject, or the creation of wealth…I define postmodern as incredulity toward metanarratives.
>
> (Lyotard 1984, pp. xxiii–xxiv)

Gary Aylesworth also describes a set of critical, strategic, and rhetorical practices of postmodernism (Aylesworth 2015).

If we regard logical positivism as a summary of modernity in philosophy of science, then all the schools in philosophy of science after logical positivism can be regarded as "postmodernism philosophy." If this is the case, the more accurate name would probably be "post-positivist" philosophy. In this sense, social constructivism and feminism can be classified as postmodernism.

However, "postmodernism" usually refers to philosophical discussions from the perspective of cultural criticism or rhetoric, such as deconstructionism and postmodern rhetoric. Representatives include Jacques Derrida, Jacques Lacan, Bruno Latour, Sara Aronowitz, and Donna Haraway, among others. In the United States, the Center for Interdisciplinary Studies in Science and Critical Theory at Duke University and the Center for the Critical Analysis of Contemporary Culture (CCACC) – now the Center for Cultural Analysis – at Rutgers University have especially done a great deal of research in a postmodern mode.

Postmodernists offer a thorough criticism of the traditional philosophy of science, and they also introduce many new concepts, such as "narrative" and "metaphor." But their deconstruction or criticism of scientific research as the function of "power" or "authority," as well as their slightly exaggerative and fashionable writing style, have aroused great dissatisfaction among scientists.

In 1996, Alan Sokal, a theoretical physicist at New York University, deliberately prepared an article, "Transforming the Boundaries: Towards Transformative Hermeneutics of Quantum Gravity," and submitted it to *Social Text*, a pioneering journal of cultural criticism. In this paper, Sokal intentionally borrowed many terms from postmodern theorists and cultural critics to argue that the reality of modern scientific research is indeed the result of social construction, just as postmodernists would claim. *Social Text* believed in the article and published it in issues 46–47 of 1996. Later, Sokal exposed the matter and criticized the poor scientific quality of postmodernist philosophers and the confusion of expert evaluation standards.

The Sokal affair thus triggered a heated debate between scientists and postmodernist philosophers, which also came to be called "Science Wars." The main works criticizing postmodernist philosophy are *Higher Superstition* by P. Gross and N. Levitt and *Fashionable Nonsense* by A. Sokal and J. Bricmont. In this debate, criticisms of postmodernism have mainly come from scientists, not from the field of philosophy of science – for example, Gross is a biologist, Levitt is a mathematician, and Sokal is a physicist. In this way, Science Wars served also to highlight the gap between science and philosophy of science. In the period of logical positivism, most philosophers of science were also scientists. Many of their works, such as Carnap's *Philosophical Foundations of Physics* and Hempel's *Philosophy of Natural Science*, were likewise well received by scientists. But in the Science Wars, many scientists and some philosophers of science stood in opposing camps. Therefore, many orthodox philosophers of science, such as Koertge, had to admit that philosophy of science had been under great external pressure and that more work would be required to eliminate the gap between science and philosophy of science (Koertge 2000).

In a paper titled "Postmodernity in Science and Philosophy," Tian-yu Cao assesses the influence of Kuhn on postmodern philosophy. Firstly, there was Quine's criticism of logical positivism, which was strengthened by Wittgenstein's self-criticism and reflection on language, and taken up by

Kuhn. Second, there was Kuhn's concept of a paradigm. As the result of historical development and the premise of scientific inquiry, this served to further introduce elements of historical and conceptual relativity into scientific understanding. In Cao's view, many points of thought in postmodernism – such as there being no knowledge as representation and there being no science with universal logic and objective truth – can be linked to Kuhn's criticism of modernity (Cao 2002, pp. 12–13).

By analyzing the three thoughts in the "new age" philosophies of science, we can see that Kuhn's relativism has played an important role. In addition, many central issues of philosophy of science, such as scientific growth, demarcation between science and pseudoscience, debates on scientific realism, science and value, and so on, all involve Kuhn's relativism. In addition, theory and observation, confirmation and acceptance, the Duhem–Quine thesis, underdetermination, the problem of induction, and prediction and evidence are more or less related to Kuhn, or at least promoted by the historical turn in philosophy of science.

It is for this reason that John Earman, a former president of the Philosophy of Science Association and an emeritus professor in the Department of History and Philosophy of Science at the University of Pittsburgh, calls himself "a distant student of Carnap and a close student of Kuhn" (Earman 1993, p. 32). One of the more recently published textbooks mentioned above, *Philosophy of Science: Central Issues*, edited by Martin Curd, is also dedicated to Hempel and Kuhn.

Summary: A personal point of view

Since philosophy of science has entered the era of post-positivism, the rise of relativism following Kuhn's concept of a paradigm has become an urgent problem to be solved. The three thoughts in the "new age" are closely related to Kuhn's historicism and more or less presuppose the problem of relativism.

Relativism is not only limited to academic discussion in the field of philosophy, but also has a significant social impact on our world. In 1993, Samuel P. Huntington, a professor at Harvard University and consultant on matters of U.S. foreign and military policy at the time, put forward the theory of "clash of civilizations." On this theory, he wrote that with the end of the Cold War, conflicts among civilizations would replace ideology and become the main line of world struggle (Huntington 1993, pp. 22–49). The September 11 attacks and the subsequent U.S. war in Afghanistan provided a prominent footnote to this theory of the "clash of civilizations."

In contrast, Hans Kung proposed a "global ethics" at the 1993 Parliament of the World's Religions in Chicago and in the 1997 Universal Declaration of Human Responsibilities, respectively, hoping to find universally applicable value standards for different cultural and ethical systems. We might say, however, that the prediction of a clash of civilizations and the search for a global ethics are different responses to the same problem: The former thinks that the differences between civilizations cannot be eliminated and will thus eventually manifest in the form of conflicts, while the latter also recognizes the differences between civilizations, but for the sake of harmonious coexistence among civilizations, highlights the need to establish universal value standards. The focus of the dispute, therefore, is ultimately whether we can overcome the relativism caused by the differences between civilizations.

Because relativists may argue that the languages and values of various civilizations are so different that we can neither achieve complete communication across civilizations, nor establish values that are universally applicable, it may seem like a natural conclusion that the differences between civilizations cannot be eliminated through rational discussion, but only in the form of conflict. Seen in this light, relativism has surely become one of the most serious challenges on the horizon since the September 11 attacks highlighted a seemingly imminent clash of civilizations. As prominent a philosophical voice as Alasdair MacIntyre speaks to the powerful and persistent appeal of relativism in our day:

> For relativism, like scepticism, is one of those doctrines that have by now been refuted a number of times too often. Nothing is perhaps a surer sign that a doctrine embodies some not to be neglected truth than that in the course of the history of philosophy it should have been refuted again and again. Genuinely refutable doctrines only need to be refuted once.
>
> (MacIntyre 1987, p. 385)

As early as in ancient Greece, Protagoras, who is best known for his proclamation "Man is the measure of all things," was claiming that human beings could only have "opinions" but no "knowledge," which was perhaps the earliest form of relativism we have on record. Plato and Aristotle both gave their own answers to this, and their philosophical systems have been regarded as the crowning achievements of ancient Greek philosophy and treasures of Western philosophy.

After the scientific revolution, Hume then came to doubt the objectivity and rationality of knowledge again from the perspective of skepticism. In *Critique of Pure Reason*, Kant answers, "How are synthetic judgements *a priori* possible?", the traditional translation of which was "How are synthetic *a priori* judgements possible?" which brought about the Copernican Revolution in philosophy. Yet Kuhn's relativism is also a challenge to Kant's philosophy. If Kant's core concept of a "category" is not universal and necessary, but rather dependent on language, and various paradigms have different languages, then our synthetic judgments – i.e., empirical sciences – are no longer *a priori*, and the universality and necessity of our scientific knowledge becomes problematic. So Kuhn later developed his so-called post-Darwinian Kantianism (Kuhn 2000, p. 104). Kuhn acknowledged Kant's famous saying

"The mind is the lawgiver of nature," but he also held that various paradigms may have different laws, so our knowledge is still relative.

In subsequent decades, there have been some criticisms of relativism in Western philosophy of science, raised by figures such as Larry Laudan, Hilary Putnam, and William Newton-Smith. However, their research has been too limited to philosophy of science, failing to touch on the deeper foundation of relativism. Thus, as of yet, there is no convincing solution. As a result, other philosophers, such as Paul Feyerabend, simply accepted relativism.

Relativism involves many philosophical concepts, such as paradigm, language, communication, incommensurability, rationality, objectivity, etc., but while relativism follows traditional philosophy's understanding of these concepts, its further criticism of traditional philosophy leads to the denial of "rationality" and "objectivity." Therefore, if we really want to overcome relativism, we must start from the root, providing new understandings for these basic concepts, so as to provide an account that adequately defends rationality and objectivity (Wang 2003).

Philosophy of science has inherited the precious heritage of logical positivism, logical analysis, but the work of logical analysis alone is far from enough. The reason why analytical philosophy has seen great achievements in recent years is that traditional speculative philosophy simply provided much material for criticism. On this basis, analytic philosophy clarifies many concepts of traditional philosophy and refines argumentation of speculative philosophy in detail. But the constructive work is still necessary. Indeed, if philosophers of science wish to provide better understandings of such various concepts as paradigm, language, rationality, and objectivity, then criticism alone will not be enough. Positive and constructive answers must also be given.

The author is personally partial to the subtitle of Michael Polanyi's book *Personal Knowledge*: "Towards a Post-Critical Philosophy." Philosophy of science has entered the post-positivist era, and the problem of relativism will force it to enter further the realm of post-critical philosophy. Although we cannot expect to build an eternal Hegelian system, we can tentatively give a comprehensive answer and then wait for later generations to criticize it, so that they can continue to move forward from this foundation.

Therefore, the author's personal view is that relativism is the most important problem confronting Western philosophy of science. Because the overcoming of relativism is no simple task, but rather an immense project comparable to Plato's and Kant's philosophy, it will require the integration of analytical philosophy and continental philosophy, along with philosophy of science, of mind, of language, and other disciplines. Imagined this way, such a project will not only summarize and criticize the traditional philosophy of science, but it will also help us better understand the new age of philosophy of science and move forward from there.

Due to the problem of relativism, philosophy of science is bound to further strengthen its relationship with philosophy of mind, philosophy of language, and sociology, and thus move more toward post-critical philosophy. If the problem of relativism can be thoroughly resolved, it will be a truly great breakthrough in the history of world philosophy. The hope is that those of us within China's philosophical circles can also make our own contributions to this difficult but glorious task.

Bibliography

Achinstein, P. (1983). *The Nature of Explanation*. New York: Oxford University Press.

Ayer, A. J. (1956). What is a Law of Nature? *Revue Internationale de Philosophie*, 10: 144–165.

Ayer, A. J. (1973). *The Central Questions of Philosophy*. London: Penguin.

Ayer, A. J. (1982). *Philosophy in the Twentieth Century*. New York: Random House.

Aylesworth, G. (2015). Postmodernism. *The Stanford Encyclopedia of Philosophy*. (Spring 2015 Edition). Edward N. Zalta (Ed.). Retrieved from https://plato.stanford.edu/archives/spr2015/entries/postmodernism/.

Bacchus, F. (1990). *Representing and Reasoning with Probabilistic Knowledge*. Cambridge, MA: MIT Press.

Baynes, K. et al. (Eds.) (1987). *After Philosophy*. Cambridge, MA: MIT Press.

Bernstein, R. J. (1988). *Beyond Objectivism and Relativism*. Philadelphia, PA: The University of Pennsylvania Press.

Bloor, D. (1976). *Knowledge and Social Imagery*. Chicago: The University of Chicago Press.

Boyd, R. et al. (Eds.) (1991). *The Philosophy of Science*. Cambridge, MA: MIT Press.

Braybrooke, D. (1987). *Philosophy of Social Science*. Englewood Cliffs, NJ: Prentice-Hall.

Brigandt, I. and Love, A. (2017). Reductionism in Biology. *The Stanford Encyclopedia of Philosophy*. (Spring 2017 Edition), Edward N. Zalta (Ed.). Retrieved from https://plato.stanford.edu/archives/spr2017/entries/reduction-biology/.

Bromberger, S. (1966). Why-questions. In: R. G. Colodny (Ed.), *Mind and Cosmos*. Pittsburgh, PA: University of Pittsburgh Press.

Brown, H. I. (1988). *Rationality*. London: Routledge.

Brown, H. I. (1990). Prospective Realism. *Studies in the History and Philosophy of Science*, 211–242.

Bunge, M. (1984). What Is Pseudoscience? *Skeptical Inquirer*, 9(1).

Cao, Tian-yu 曹天予. (2002). 科学和哲学中的后现代性．曹南燕译．哲学研究, 2.

Carnap, R. (1932). The Elimination of Metaphysics Through Logical Analysis of Language. *Erkenntnis*, 2: 60–81.

Carnap, R. (1950). *Logical Foundation of Probability*. Chicago: The University of Chicago Press.

Carnap, R. (1966). *Philosophical Foundations of Physics: An Introduction to the Philosophy of Science*. New York: Basic Books.

Carnap, R. (1995). *An Introduction to the Philosophy of Science*. M. Gardner (Ed.). New York: Dover.

Cartwright, N. (1983). *How the Laws of Physics Lie*. Oxford: Clarendon.

Cartwright, N. (1999). *The Dappled World*. Cambridge: Cambridge University Press.

Chalmers, A. F. (1999). *What Is This Thing Called Science?* (3rd Edition). Brisbane: University of Queensland Press.

Chen, Bo 陈波. (2001). 休谟问题和金岳霖的回答．中国社会科学，3.

Chen, Jiang (1997). 陈健. 科学划界．北京：东方出版社.

Chen, Xiao-ping 陈晓平. (1994). 归纳逻辑与归纳悖论. 武汉大学出版社.

Chen, Xiao-ping 陈晓平. (1997). 大弃赌定理及其哲学意蕴．自然辩证法通讯，(2).

Churchland, P. M. and Hooker, C. A. (Eds.) (1985). *Images of Science*. Chicago: The University of Chicago Press.

Cohen, J. (1989). *An Introduction to the Philosophy of Induction and Probability*. Oxford: Clarendon Press.

Cohen, J. and Hesse, M. (Eds.) (1980). *Application of Inductive Logic*. Oxford: Clarendon Press.

Corfield, D. and Williamson, J. (2001). *Foundations of Bayesianism*. Dordrecht: Kluwer Academic Publishers.

Couvalis, G. (1997). *The Philosophy of Science: Science and Objectivity*. London: Sage.

Craig, E. (Ed.) (2000). *Routledge Encyclopedia of Philosophy* CD-ROM. London: Routledge.

Curd, M. and Cover, J. A. (Eds.) (1998). *Philosophy of Science: The Central Issues*. New York: W. W. Norton & Company.

Dray, W. H. (1964). *Philosophy of History*. Englewood Cliffs, NJ: Prentice-Hall.

Dretske, F. I. (1977). Laws of Nature. *Philosophy of Science*, 44: 248–268.

Dubucs, J. (Ed.) (1993). *Philosophy of Probability*. Dordrecht: Kluwer Academic Publishers.

Earman, J. (1992). *Bayes or Bust? A Critical Examination of Bayesian Confirmation Theory*. Cambridge, MA: MIT Press.

Earman, J. (1993). Carnap, Kuhn, and the Philosophy of Scientific Methodology. In: P. Horwich (Ed.), *World Changes*. Cambridge, MA: MIT Press.

Earman, J. et al. (Eds.) (2002). *Ceteris Paribus Laws*. Norwell, MA and Dordrecht: Kluwer Academic Publishers.

Einstein, A. (1961). *Relativity*. R. W. Lawson (Trans.). New York: Three Rivers Press.

Einstein, A. (2013). *Albert Einstein, The Human Side: Glimpses from His Archives*. Helen Dukas et al. (Eds.). Princeton, NJ: Princeton University Press.

Ellis, B. (1979). *Rational Belief Systems*. Oxford: Blackwell.

Fan, Dai-nian 范岱年. (1998). 一部为科学实在论作辩护的当代物理学思想史．自然辩证法研究，1.

Feigl, H. and Maxwell, G. (Eds.) (1962). *Scientific Explanation, Space, and Time*. Minneapolis, MN: University of Minnesota Press.

Fetzer, J. (2017). Carl Hempel. *The Stanford Encyclopedia of Philosophy*. (Fall 2017 Edition), Edward N. Zalta (Ed.). Retrieved from https://plato.stanford.edu/archives/fall2017/entries/hempel/.

Fetzer, J. (Ed.) (2000). *Science, Explanation, and Rationality: The Philosophy of Carl G. Hempel*. Oxford and New York: Oxford University Press.

Feyerabend, P. (1975). *Against Method*. London: Verso.

Feyerabend, P. (1998). How to Defend Society against Science. In: E. D. Klemke et al. (Eds.), *Introductory Readings in the Philosophy of Science* (3rd Edition). New York: Prometheus Books.

Fine, A. (1984). The Natural Ontological Attitude. In: J. Leplin (Ed.), *Scientific Realism*. Berkeley: University of California Press.

Fine, A. (2000). Scientific Realism and Antirealism. In: E. Craig (Ed.), *Routledge Encyclopedia of Philosophy* CD-ROM. London: Routledge.

Franklin, A. (1998). Do Mutants Die of Natural Causes? The Case of Atomic Parity Violation. In: N. Koertge (Ed.), *A House Built on Sand: Exposing Postmodernist Myths about Science*. Oxford: Oxford University Press.

Friedman, M. (1974). Explanation and Scientific Understanding. *The Journal of Philosophy*, 71: 5–19.

Galison, P. (1987). *How Experiments End*. Chicago: The University of Chicago Press.

Gibson, Q. (1960). *The Logic of Social Enquiry*. London: Routledge & K. Paul.

Giere, R. (1999). *Science without Laws*. Chicago: The University of Chicago Press.

Glock, H. (1996). *A Wittgenstein Dictionary*. Oxford: Blackwell Publisher Ltd.

Glymour, C. (1980). *Theory and Evidence*. Princeton, NJ: Princeton University Press.

Goldman, R. N. (1997). *Einstein's God—Albert Einstein's Quest as a Scientist and as a Jew to Replace a Forsaken God*. Northvale, NJ: Joyce Aronson Inc.

Goodman, N. (1983). *Fact, Fiction, and Forecast* (4th Edition). Cambridge, MA: Harvard University Press.

Gross, P. and Levitt, N. (1994). *Higher Superstition: The Academic Left and its Quarrels with Science*. Baltimore, MD: The Johns Hopkins Press.

Guo, Gui-chun 郭贵春编. (2000). 走向21世纪的科学哲学．太原：山西科学技术出版社.

Guo, Gui-chun 郭贵春. (2001). 科学实在论教程．北京：高教出版社.

Habermas, J. (1971). *Knowledge and Human Interests*. Boston: Beacon Press.

Hacking, I. (1982). Experimentation and Scientific Realism. *Philosophical Topics*, 13: 71–87.

Hacking, I. (1983). *Representing and Intervening*. Cambridge: Cambridge University Press.

Hacking, I. (1999). *The Social Construction of What?* Cambridge, MA: Harvard University Press.

Hanson, N. R. (1958). Patterns of Discovery. Cambridge: Cambridge University Press.

He, Zuo-xiu 何祚庥编. (1996). 伪科学曝光．北京：中国社会科学出版社.

He, Zuo-xiu 何祚庥编. (1999). 伪科学再曝光．北京：中国社会科学出版社.

Hegel, G. W. F. (1956). *The Philosophy of History*. J. Sibree (Trans.). New York: Dover Publications.

Hempel, C. G. (1950). Problems and Changes in the Empiricist Criterion of Meaning. *Revue Internationale de Philosophie*, 4(11): 41–63.

Hempel, C. G. (1965). *Aspects of Scientific Explanation and Other Essays in the Philosophy of Science*. New York: The Free Press.

Hempel, C. G. (1966). *Philosophy of Natural Science*. Englewood Cliffs, NJ: Prentice-Hall.

Hempel, C. G. (1968). Explanation in Science and in History. In P. H. Nidditch (Ed.). The *P*hilosophy of *S*cience. London: Oxford University Press.

Hempel, C. G. (1998). Science and Human Value. In: E. D. Klemke et al. (Eds.), *Introductory Readings in the Philosophy of Science* (3rd Edition). New York: Prometheus Books.

Hempel, C. G. (2001). *The Philosophy of Carl G. Hempel: Studies in Science, Explanation, and Rationality*. J. H. Fetzer (Ed.). Oxford and New York: Oxford University Press.

Henderson, L. (2018). Problem of Induction. *The Stanford Encyclopedia of Philosophy*. (Spring 2020 Edition). Edward N. Zalta (Ed.). Retrieved from https://plato.stanford.edu/archives/spr2020/entries/induction-problem/.

Hilpinen, R. (Ed.) (1980). *Rationality in Science*. Dordrecht: D. Reidel Publishing.

Ho, Hsiu-hwang 何秀煌. (1999). 记号、意识与典范．台北：东大图书公司.

Hollinger, R. (1998). From Weber to Habermas. In: E. D. Klemke et al. (Eds.), *Introductory Readings in the Philosophy of Science* (3rd Edition). New York: Prometheus Books.

Hollis, M. and Lukes S. (Eds.) (1982). *Rationality and Relativism*. Oxford: Basil Blackwell.

Hong, Qian 洪谦. (1999). 论逻辑经验主义．北京：商务印书馆.

Hong, Qian 洪谦. (1990). 逻辑经验主义论文集．三联书店(香港)有限公司.

Hong, Qian 洪谦. (1989). 维也纳学派哲学．北京：商务印书馆.

Horwich, P. (1990). *Truth*. Oxford: Basil Blackwell.

Horwich, P. (Ed.) (1993). *World Change*. Cambridge, MA: MIT Press.

Howson, C. and Urbach, P. M. (1993). *Scientific Reasoning: The Bayesian Approach* (2nd Edition). Chicago: Open Court.

Huang,Shun-ji 黄顺基等编.(1991).科学技术哲学的前沿与进展．北京：人民出版社.

Huang, Shun-ji and Liu, Da-chun 黄顺基、刘大椿 主编 (Eds.) (1991). 科学技术哲学的前沿与进展. 北京: 人民出版社.

Hull, D. (1988). *Science as a Process: An Evolutionary Account of the Social and Conceptual Development of Science*. Chicago: The University of Chicago Press.

Hume, David (2007a). *A Treatise of Human Nature*. D. F. Norton and M. J. Norton (Eds.). Oxford: Oxford University Press.

Hume, David (2007b). *An Enquiry Concerning Human Understanding*. P. Millican (Ed.). Oxford: Oxford University Press.

Huntington, S. P. (1993). The Clash of Civilizations. *Foreign Affairs*, 72(3).

Jeffreys, H. (1961). *Theory of Probability* (3rd Edition). Oxford: Clarendon Press.

Jiang, Tian-ji 江天骥. (1984). 当代西方科学哲学．北京：中国社会科学出版社.

Jiang, Tian-ji 江天骥编. (1988). 科学哲学名著选读．武汉：湖北人民出版社.

Jiang, Yi. (1998). 江怡.维特根斯坦传.石家庄：河北人民出版社.

Jin, Wu-lun 金吾伦. (1994). 托马斯・库恩．三联书店(香港)有限公司.

Kant, I. (1929). *Critique of Pure Reason*. N. K. Smith (Trans.). New York: MacMillan.

Keller, E. F. (1983). *A Feeling for the Organism: The Life and Work of Barbara McClintock*. San Francisco: W. H. Freeman.

Kitcher, P. (1981). Explanatory Unification. *Philosophy of Science*, 48: 507–53.

Kitcher, P. (1993). *The Advancement of Science: Science without Legend, Objectivity without Illusions*. Oxford: Oxford University Press.

Klee, R. (1999). *Scientific Inquiry: Readings in Philosophy of Science*. Oxford: Oxford University Press.

Klemke, E. D. et al. (Eds.) (1998). *Introductory Readings in the Philosophy of Science* (3rd Edition). New York: Prometheus Books.

Koertge, N. (2000). 'New Age' Philosophies of Science: Constructivism, Feminism and Postmodernism. *British Journal for the Philosophy of Science*, 51: 667–683.

Koyre, A. (1978). *Galileo Studies*. Atlantic Highland, NJ: Humanities Press.

Kuhn, T. (1970). *The Structure of Scientific Revolutions* (2nd Edition). Chicago: The University of Chicago Press.

Kuhn, T. (1977). *The Essential Tension*. Chicago: The University of Chicago Press.

Kuhn, T. (2000). *The Road since Structure*. J. Conant and J. Haugeland (Eds.). Chicago: The University of Chicago Press.
Lakatos, I. (1978). *The Methodology of Scientific Research Programmes*. J. Worrall and G. Currie (Eds.). Cambridge: Cambridge University Press.
Lakatos, I. and Musgrave, A. (Eds.) (1970). *Criticism and the Growth of Knowledge*. Cambridge: Cambridge University Press.
Latour, B. and Woolgar, S. (1979). *Laboratory Life: The Social Construction of Scientific Facts*. Princeton, NJ: Princeton University.
Lau, Din-cheuk (Trans.) (2003). *Mencius*. Hong Kong: The Chinese University Press.
Laudan, L. (1977). *Progress and Its Problems*. London: Routledge & K. Paul.
Laudan, L. (1981). A Confutation of Convergent Realism. *Philosophy of Science*, 48: 19–49.
Laudan, L. (1984). *Science and Value*. Berkeley: University of California Press.
Laudan, L. (1990). *Science and Relativism*. Chicago: The University of Chicago Press.
Laudan, L. (1996). *Beyond Positivism and Relativism*. Boulder, CO: Westview Press.
Leplin, J. (Ed.) (1984). *Scientific Realism*. Berkeley: University of California Press.
Lewis, D. (1973). *Counterfactuals*. Cambridge, MA: Harvard University Press.
Lin, Cheng-hung 林正弘. (1989). 伽利略、波柏、科学说明．台北： 东大图书股份有限公司.
Liu, Chuang (1997). Models and Theories I: The Semantic View Revisited. *International Studies in the Philosophy of Science*, 11(2).
Liu, Da-chun 刘大椿编. (1998). 科学哲学通论．北京：中国人民大学出版社.
Liu, Da-chun 刘大椿. (1998). 科学哲学．北京：人民出版社.
Liu, Da-chun 刘大椿. (2000). 科学技术哲学导论．北京：中国人民大学出版社.
Liu, Hua-jie 刘华杰. (2000). 科学元勘中SSK学派的历史与方法论述评．哲学研究， 1.
Lyotard, J. (1984). *The Postmodern Condition: A Report on Knowledge*. G. Bennington and B. Massumi (Trans.). Minneapolis, MN: Minnesota University Press.
MacIntyre, A. (1987). Relativism, Power and Philosophy. In: K. Baynes et al. (Eds.), *After Philosophy*. Cambridge, MA: MIT Press.
MacKenzie, D. (1978). Statistical Theory and Social Interests: A Case Study. S*ocial Studies of Science*, 8.
Masterman, M. (1970). The Nature of a Paradigm. In: I. Lakatos and A. Musgrave (Eds.), *Criticism and the Growth of Knowledge*. Cambridge: Cambridge University Press.
Maxwell, G. (1962). The Ontological Status of Theoretical Entities. In: H. Feigl and G. Maxwell (Eds.), *Scientific Explanation, Space, and Time. Minnesota Studies in the Philosophy of Science* (Vol. 3). Minneapolis, MN: University of Minnesota Press.
McCarthy, T. (1977). On an Aristotelian Model of Scientific Explanation. *Philosophy of Science*, 44(1): 159–166.
McErlean, J. (2000). *Philosophies of Science: From Foundations to Contemporary Issues*. Belmont, CA: Wadsworth Publishing Co.
McMullin, E. (1993). Rationality and Paradigm Change in Science. In: P. Horwich (Ed.), *World Changes*. Cambridge, MA: MIT Press.
Meiland, J. W. and Krausz, M. (Eds.) (1982). *Relativism*. Notre Dame, IN: University of Notre Dame Press.
Mill, J. S. (1904). *A System of Logic*. New York: Harper and Row.
Mitchell, S. (2000). Dimensions of Scientific Law. *Philosophy of Science*, 67(2): 242–265.

Morton, A. (2000). Saving Epistemology from the Epistemologists: Recent Work in the Theory of Knowledge. *British Journal for the Philosophy of Science*, 51(4): 685–704.

Musgrave, A. (1985). Realism versus Constructive Empiricism. In: P. M. Churchland and C. A. Hooker (Eds.), *Images of Science*. Chicago: The University of Chicago Press.

Musgrave, A. (1989). NOA's Ark—Fine for Realism. *Philosophical Quarterly*, 39: 383–398.

Nagel, E. (1961). *The Structure of Science: Problems in the Logic of Scientific Explanation*. London: Routledge & K. Paul.

Nagel, E. and Brandt, R. B. (Eds.) (1965). *Meaning and Knowledge*. San Diego: Harcourt, Brace & World.

Newton-Smith, W. H. (1981). *The Rationality of Science*. London: Routledge.

Nidditch, P. H. (Ed.) (1968). *The Philosophy of Science*. London: Oxford University Press.

Nozick, R. (1993). *The Nature of Rationality*. Princeton, NJ: Princeton University Press.

Nye, A. (Ed.) (1998). *Philosophy of Language: The Big Questions*. Malden, MA: Blackwell.

Palmer, P. E. (1969). *Hermeneutics*. Evanston, IL: Northwestern University Press.

Pickering, A. (1984). *Constructing Quarks: A Sociological History of Particle Physics*. Chicago: The University of Chicago Press.

Polanyi, M. (1958). *Personal Knowledge: Towards A Post-Critical Philosophy*. Chicago: The University of Chicago Press.

Popper, K. (1959). *The Logic of Scientific Discovery*. London: Hutchinson.

Popper, K. (1962). *Conjectures and Refutations: The Growth of Scientific Knowledge*. New York: Basic Books.

Popper, K. (1974). Autobiography of Karl Popper. In: P. Schilpp (Ed.), *The Philosophy of Karl Popper*. Chicago: Open Court.

Popper, K. (2013). *The Poverty of Historicism* (2nd Edition). London and New York: Routledge.

Pryor, J. (2000). Highlights of Recent Epistemology. *British Journal for the Philosophy of Science*, 51: 685–704.

Psillos, S. (2000). The Present State of the Scientific Realism Debate. *British Journal for the Philosophy of Science*, 51: 705–728.

Psillos, S. (2002). *Causation and Explanation*. Montreal: McGill-Queen's University Press.

Putnam, H. (1975). *Mathematics, Matter and Method*. Cambridge: Cambridge University Press.

Putnam, H. (1981). *Reason, Truth and History*. Cambridge: Cambridge University Press.

Putnam, H. (1982). Three Kinds of Scientific Realism. *The Philosophical Quarterly*, 32(128): 195–200.

Quine, W. V. O. (1974). On Popper's Negative Methodology. In: P. A. Schlipp (Ed.), *The Philosophy of Karl Popper*. Chicago: Open Court.

Quine, W. V. O. (1980). *From a Logical Point of View*. Cambridge, MA: Harvard University Press.

Quine, W. V. O. (1998). Indeterminacy of Translation. In: A. Nye (Ed.), *Philosophy of Language: The Big Questions*. Malden, MA: Blackwell.

Radder, H. (Ed.) (2003). *The Philosophy of Scientific Experimentation*. Pittsburgh, PA: University of Pittsburgh Press.

Railton, P. (1978). A Deductive-nomological Model of Probabilistic Explanation. Philosophy of Science, 45, 206–226.

Ramsey, F. (1978). *Foundation of Mathematics*. Atlantic Highlands, NJ: Humanities Press.

Reichenbach, H. (1951). *The Rise of Scientific Philosophy*. Berkeley-Los Angeles: University of California Press.

Rorty, R. (1991). *Objectivity, Relativism, and Truth*. Cambridge: Cambridge University Press.

Rosser, S. (1990). *Female-Friendly Science: Applying Women's Studies Methods and Theories to Attract Students*. New York: Teachers College Press.

Rossiter, M. (1985). *Women Scientists in America: Struggles and Strategies to 1940*. Baltimore: The Johns Hopkins Press.

Ruben, D. (1990). *Explaining Explanation*. New York: Routledge.

Rudner, R. (1998). The Scientist *Qua* Scientist Makes Value Judgments. In: E. D. Klemke et al. (Eds.), *Introductory Readings in the Philosophy of Science* (3rd Edition). New York: Prometheus Books.

Ruse, M. (1995). Auguste Comte. In: T. Honderich (Ed.), *The Oxford Companion to Philosophy*. Oxford: Oxford University Press.

Salmon, W. (1984). *Logic* (3rd Edition). Upper Saddle River, NJ: Prentice-Hall.

Salmon, W. (1989). *Four Decades of Scientific Explanation*. Minneapolis, MN: University of Minnesota Press.

Salmon, W. (1990). Rationality and Objectivity in Science or Tom Kuhn Meets Tom Bayes. In: C. Wade Savage (Ed.), *Scientific Theories*. Minneapolis, MN: University of Minnesota Press.

Salmon, W. (1998). *Causality and Explanation*. Oxford: Oxford University Press.

Sankey, H. (1997). Rationality, Relativism and Incommensurability. Farnham, UK: Ashgate Publishing.

Sarkar, S. (Ed.) (1996). *The Emergence of Logical Empiricism: From 1900 to the Vienna Circle*. New York: Garland Publishing.

Scheffler, I. (1964). *The Anatomy of Inquiry*. London: Routledge & K. Paul Ltd.

Scheffler, I. (1967). *Science and Subjectivity*. Bobbs-Merrill

Schiebinger, L. (1999). *Has Feminism Changed Science?* Cambridge, MA: Harvard University Press.

Schilpp, P. A. (Ed.) (1963). *The Philosophy of Rudolf Carnap*. Chicago: Open Court

Schilpp, P. A. (Ed.) (1974). *The Philosophy of Karl Popper*. Chicago: Open Court.

Schlick, M. (1936). Meaning and Verification. *The Philosophical Review*, 45(4).

Schlick, M. (1959). The Turning Point in Philosophy. In: A. J. Ayer (Ed.), *Logical Positivism*. Glencoe: Free Press.

Scriven, M. (1959). Definitions, Explanations, and Theories. In: H. Feigl, M. Scriven, and G. Maxwell (Eds.), *Minnesota Studies in the Philosophy of Science* (Vol. 2). Minneapolis: University of Minnesota Press.

Scriven, M. (1962). Explanations, Predictions, and Laws. In: H. Feigl and G. Maxwell (Eds.), *Minnesota Studies in the Philosophy of Science* (Vol. 3). Minneapolis: University of Minnesota Press.

Scriven, M. (1963). The Temporal Asymmetry Between Explanations and Predictions. In: B. Baumrin (Ed.), *Philosophy of Science: The Delaware Seminar* (Vol. 1). New York: John Wiley.

Searle, J. (1995). *The Construction of Social Reality*. New York: The Free Press.

Searle, J. (1999). *Mind, Language and Society*. London: Phoenix.

Sellars, W. (1962). *Science, Perception and Reality*. New York: Humanities Press.
Shapere, D. (1984). Reason and the Search for Knowledge: Investigations in the Philosophy of Science. Dordrecht: D. Reidel.
Shapin, S. and Schaffer, S. (1986). *Leviathan and the Air-Pump: Hobbes, Boyle, and the Experimental Life*. Princeton, NJ: Princeton University.
Sheng, Wei-tong 盛维通等编. (2000). 科学技术哲学教程．北京：中国环境科学出版社.
Shi, Yan-fei 施雁飞. (1991). 科学解释学．长沙：湖南出版社.
Shu, Wei-guang and Qiu, Ren-zong 舒炜光、邱仁宗主编. (1987). 当代西方科学哲学述评. 北京：人民出版社.
Shu, Wei-guang and Qiu, Ren-zong 舒炜光、邱仁宗主编. (2007). 当代西方科学哲学述评. 第二版. 北京：中国人民大学出版社.
Smart, J. J. C. (1968). *Between Science and Philosophy*. New York: Random House.
Sokal, A. and Bricmont, J. (1998). *Fashionable Nonsense: Postmodern Intellectuals' Abuse of Science*. London: Profile Books.
Sommers, C. (1995). *Who Stole Feminism? How Women Have Betrayed Women*. New York: Simon and Schuster.
Stove, D. C. (1982). *Popper and After: Four Modern Irrationalists*. Oxford: Pergamon Press.
Sullivan, P. (1998). An Engineer Dissects Two Case Studies: Hayles on Fluid Mechanics, and MacKenzie on Statistics. In: N. Koertge (Ed.), *A House Built on Sand: Exposing Postmodernist Myths about Science*. Oxford: Oxford University Press.
Suppe, F. (Ed.) (1977). *The Structure of Scientific Theories*. Champaign, IL: University of Illinois Press.
Talbott, W. (2008). Bayesian Epistemology. *The Stanford Encyclopedia of Philosophy*. (Winter 2016 Edition). Edward N. Zalta (Ed.). Retrieved from https://plato.stanford.edu/archives/win2016/entries/epistemology-bayesian/.
Tennant, N. (2017). Logicism and Neologicism. *The Stanford Encyclopedia of Philosophy*. (Winter 2017 Edition). Edward N. Zalta (Ed.). Retrieved from https://plato.stanford.edu/archives/win2017/entries/logicism/.
Thagard, P. (1978). Why Astrology Is a Pseudoscience. In: P. Asquith and I. Hacking (Eds.), *Proceedings of the Philosophy of Science Association*, Vol. 1. East Lansing, MI: Philosophy of Science Association.
Thagard, P. (1988). *Computational Philosophy of Science*. Cambridge, MA: MIT Press.
van Fraassen, B. (1977). The Pragmatics of Scientific Explanation. *American Philosophical Quarterly*, 1977(2), 143–150.
van Fraassen, B. C. (1980). *The Scientific Image*. Oxford: Clarendon Press.
van Fraassen B. C. (1990). *Laws and Symmetry*. Oxford: Oxford University Press.
von Neumann, J. (1955). *Mathematical Foundations of Quantum Mechanics*. Princeton, NJ: Princeton University.
Wang, Wei 王巍. (2003). 相对主义：从典范、语言和理性的观点看．北京：清华大学出版社.
Wang, Wei (2017). *Explanation, Laws, and Causation*. London and New York: Routledge.
Wang, Xianjun 王宪钧. (1982). 数理逻辑引论．北京：北京大学出版社.
Watkins, J. (1984). *Science and Skepticism*. Princeton, NJ: Princeton University Press.
Weatherford, R. (1982). *Philosophical Foundations of Probability Theory*. London: Routledge & K. Paul.
Wilson, B. R. (Ed.) (1977). *Rationality*. Oxford: Basil Blackwell.
Winch, P. (1987). *Trying to Make Sense*. Oxford: Basil Blackwell.

Winch, P. (1990). *The Idea of Social Science and Its Relation to Philosophy* (2nd Edition). London: Routledge.

Wittgenstein, L. (1961). *Tractatus Logico-Philosophicus*. D. F. Pears and B. F. McGuinness (Trans.). London and New York: Routledge.

Wittgenstein, L. (1967). *Philosophical Investigations*. G. E. M. Anscombe (Trans.). Oxford: Basil Blackwell.

Woodward, J. (2014). Scientific Explanation. *The Stanford Encyclopedia of Philosophy*. (Winter 2019 Edition). Edward N. Zalta (Ed.). Retrieved from https://plato.stanford.edu/archives/win2019/entries/scientific-explanation/.

Woolgar, S. and Ashmore, M. (1988). The Next Step: An Introduction to the Reflexive Project. In: Steve Woolgar (Ed.), *Knowledge and Reflexivity*. Thousand Oaks, CA: SAGE Publications.

Xia, Ji-song and Shen, Fei-feng 夏基松、沈斐凤. (1987). 西方科学哲学．南京：南京大学出版社.

Yin, Zheng-kun and Qiu, Ren-zong 殷正坤、邱仁宗. (1996). 科学哲学引论．武汉：华中理工大学出版社.

Zalta, E. (2019). Gottlob Frege. *The Stanford Encyclopedia of Philosophy*. (Winter 2019 Edition). Edward N. Zalta (Ed.). Retrieved from https://plato.stanford.edu/archives/win2019/entries/frege/.

Zhang, Hua-xia 张华夏.(2002). 科学解释标准模型的建立、困难与出路．科学技术与辩证法, 1.

Zuckerman, H. et al. (Eds.) (1991). *The Outer Circle: Women in the Scientific Community*. New York: W. W. Norton & Co.

Name index

Subject index